Chemistry Concepts for the Real World

Second Edition

Debra Mixon

Unit 1
Classification of Matter and Measurement

Did you know that?

Alchemists used tricks to make less precious metals look like gold (Au).

The property of density can be used to determine if a metal is actually gold.

Tungsten (W) is used in incandescent light bulbs because it has the highest melting point of the metals and can withstand high temperatures.

Air is a mixture of many different gases.

Brass, bronze, and steel are mixtures of two or more metals.

Even though milk looks like it is a solution, it is actually a mixture of milk fat particles floating in a solution.

Gallium (Ga) is a metal that will melt in your hands.

Carbon (C) is used in charcoal water filters to purify water.

Mercury (Hg) is a neurotoxin that was once used to treat the animal fur used to make hats.

Ludwig van Beethoven died from lead (Pb) poisoning.

Hydrogen (H) was once used in blimp like aircraft called Zeppelins until it was realized that it was highly flammable.

Abraham Lincoln suffered from mercury (Hg) poisoning.

Objectives for Unit 1

- I can determine the number of significant digits in a quantity and perform calculations involving significant figures.

- I can distinguish between pure substances (elements and compounds) and mixtures using word descriptions, chemical formulas, and particle diagrams.

- I can determine the independent and dependent variable in an experiment.

- I can make a proper data table.

- I can measure using equipment to the correct precision.

- I can determine the difference between measurements that are precise, accurate, and accurate and precise.

- I can distinguish between chemical and physical properties and changes.

- I can use characteristic properties to determine the identity of a pure substance.

- I can determine if a reaction or process is endothermic or exothermic based on given information.

- I can calculate percent error.

- I can calculate the density of a substance given measurements for volume and mass.

Self -Test 1.1 Physical and Chemical Changes

1. Mark the following as true or false.

    a) ____ A chemical change results in the formation of a new substance or substances.

2. Label each scenario below as a physical change or a chemical change and describe how you know.

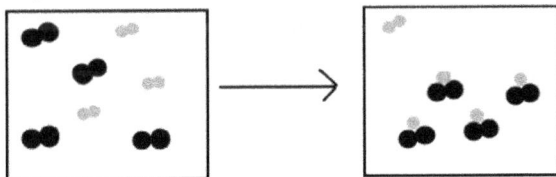 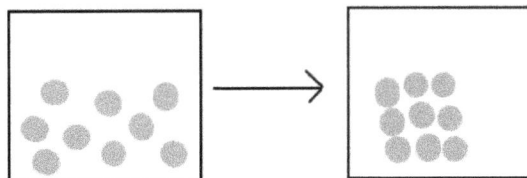

A _____   B _____

_____   _____

_____   _____

3. Mark each of the following as either a physical change (P) or a chemical change (C).

    a) ____ Sugar burning as it is heated.

    b) ____ Apples exposed to air turning brown.

    c) ____ Bleach turning hair yellow.

    d) ____ Ice melting.

    e) ____ Milk turning sour.

    f) ____ Salt dissolving.

    g) ____ Aluminum foil being crumpled.

4. Sign(s) that a chemical change has occurred include all of the following: (mark all that apply)
    a) a gas is formed

    b) a solid is formed when two solutions are mixed

    c) an explosion or flame is produced

    d) a solid disappears when placed in water

    e) a color change occurs

5. The rusting of a nail is a chemical change because:

    a) a new substance is formed.

    b) the reaction is reversible.

    c) the nail remains unchanged.

    d) the physical phase of the nail stays the same.

6. Which statement describes a physical property of oxygen?

    a) Oxygen has a density of 1.43 g/cm^3.

    b) Oxygen combines with gasoline to form carbon dioxide and water.

    c) Oxygen is used in metabolism to break down glucose.

    d) Oxygen is a gas at room temperature.

7. Mark each of the following with an "A" for endothermic or a "B" for exothermic.

    a) _____human metabolism

    b) _____ice melting

    c) _____wood burning

    d) _____an activated handwarmer

    e) _____baking soda dissolving in water causing a decrease in the water temperature

    f) _____a lake beginning to freeze in winter

Self-Test 1.2 Data Table Design

To design a proper data table, you must be able to identify the following:
   a) the **independent variable**- this is the variable that is controlled or manipulated
   b) the number of levels or conditions of the independent variable
   c) the **dependent variable** – this is the variable that is measured
   d) the number of trials of each level or condition

Example Scenario

A student hypothesized that different brands of soda contain different amounts of carbon dioxide. She designed an experiment to test the amount of carbon dioxide in a 12-oz bottle of soda. She chose to test Coke, Diet Coke, and Sprite. She tested five bottles of each type of soda and measured the amount of carbon dioxide in milliliters.

   a) The independent variable in the scenario is the type of soda.
   b) The levels of the independent variable include Coke, Diet Coke, and Sprite.
   c) The dependent variable is the amount of carbon dioxide measured in mL.
   d) The number of trials is five because there are five cans tested for each level.

(Label)   (Title)

Table 1. Brand of Soda and Amount of Carbon Dioxide

Independent Variable: Type of Soda	Dependent Variable and units: Volume of Carbon Dioxide (mL)				
	Trials				
	1	2	3	4	5
Coke					
Diet Coke					
Sprite					

(Levels or Conditions of the Independent Variable)

Your Turn:

1. **Construct a table below for the following scenario:**
A student conducted an experiment to determine if the color of light a bean plant receives affects the growth of the plant. A group of three plants of equal height were exposed to the following colors of light: green, red, and blue. Each group was placed in a separate room and only exposed to one color of light. All other variables, such as temperature, amount of water, and humidity were controlled. After two weeks, the height of each plant was measured in centimeters and recorded.

    a. What is the independent variable?_____

    b. What are the conditions or levels of the independent variable?_____

    c. What is the dependent variable?_____

    d. How many trials are conducted for each condition?_____

Fill in the table in the space below. Include a **label, title, column headings, and units**.

2. A student wished to determine the level of voltage that would plate the most copper onto a zinc disk from a copper chloride solution. She hypothesized that a higher voltage would plate the most copper. The student set up 12 copper chloride solutions, each with the same concentration and amount of solution. 12 zinc disks were obtained and the initial mass (in grams, g) of each was measured and recorded. 4 of these disks were placed in solution and attached to a power source supplying 1 volt of electricity. 4 other disks were supplied with 3 volts, and 4 disks were supplied with 6 volts. The disks were monitored for 1 hour then were removed from the solution, dried, and massed to obtain the final mass.

    a. What is the independent variable?_____

    b. What are the conditions or levels of the independent variable?_____

    c. What is the dependent variable?_____

    d. What measurements must be made and how many times must they be collected for each zinc disk?_____

    e. How many trials are conducted for each condition?_____

Fill in the table in the space below for the experiment described in example 2. Include a **label, title, column headings, and units**.

3. A student designed a class experiment to test the effect of different ionic compounds (salts) on melting ice. The student chose three different compounds (sodium chloride, calcium chloride, and sodium acetate) to test against a control that did not receive any salt. In the experiment, 50.0 grams of ice was added to 3 containers for each condition and 5.00 grams of each salt was added into the respective containers. After 3 minutes, the amount of melt water produced in each container was measured in mL.

    a) Design a data table for the experiment in the space below. Include a **label, title, column headings, and units**.

Name_____

Self-Test 1.3 Pure Substances, Elements, Compounds, and Mixtures

1. Fill in the following table regarding pure substances:

Type of Matter	How many kinds of **atoms** make up this type of matter?	This type of matter can be broken down into simpler substances by physical or chemical means. (T/F)	Write an example	
			Name	Formula
Element				
Compound				

2. Using the following symbols, Ag = **O**, Cl₂ = ◆◆ , and H₂O = 🐭 , fill in the table with the appropriate <u>drawings</u>.

10 atoms of an element	5 molecules of an element	5 molecules of a compound

3. Classify the following as element (E), compound (C), or mixture (M).

Substance	E, C, or M	Substance	E, C, or M
saltwater		silver plated jewelry	
copper wire		Sprite™	
rust (Fe_2O_3)		air	
granite		Ice water	

4. The middle box below shows a substance in the liquid phase. Fill in the box on the left with how the substance would look in the **solid** phase, and the box on the right with how it would look in the **gas** phase. Label the arrows with the appropriate physical change (boiling, freezing, melting, condensing).

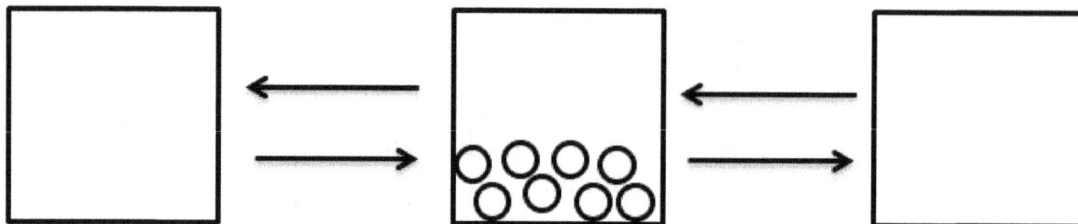

5. Which state, solid, liquid, or gas, is associated with the greatest kinetic energy?_____

6. For each of the chemical formulas below, fill in the table with the name and number of atoms of each element in the substance.

Substance	Element Name	# of Atoms	Element Name	# of Atoms	Element Name	# of Atoms
$CaCl_2$						
$C_6H_{12}O_6$						
$CuSO_4$						
Br_2						

7. Which substance in the table above is not a compound?_____

8. Mark each of the boxes below as containing one or more of the following: mixture, pure substance, gas, solid, liquid, element, compound. One of the boxes contains 5 of these; which one?_____

a)   b)   c)   d)   e)

Self-Test 1.4 Characteristic Properties

1. An experiment was done to determine the relative densities of two liquids and one solid. The results are seen below. Indicate the relative densities.

    a) Most dense _____

    b) Middle density _____

    c) Least dense _____

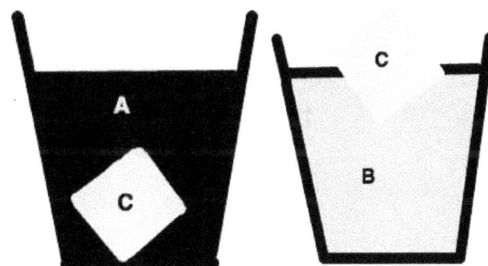

2. Two balls are made of the same material. The larger of the two balls is put into water with the result shown on the left. Which picture on the right shows the result when the smaller ball is placed in the water?

  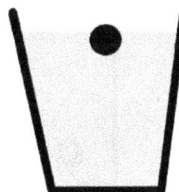 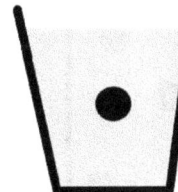 

Large Ball   A   B   C   D

3. Sample 1 of pure aluminum (Al) has a mass of 10.0 grams while sample 2 of pure aluminum has a mass of 100.0 grams. For each property listed below, write
    - "A" if sample 1 is less than sample 2,
    - "B" if sample 1 is greater than sample 2,
    - "C" if sample 1 is equal to sample 2.

Property	A, B, or C
Melting Point	
Boiling Point	
Volume	
Density	
Mass	

4. In an experiment, two pure substances are compared to determine if they are made of the same material. Which of the following properties can be measured to provide evidence that the substances are the same. Mark all that apply.

    ____Melting point

    ____Boiling point

    ____Color

    ____Physical state

    ____Mass

    ____Density

    ____Smell

    ____Volume

5. Circle the state of matter that is the least dense.

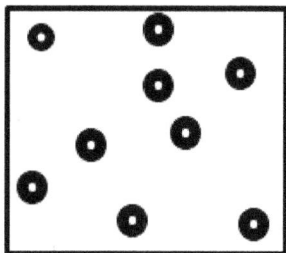

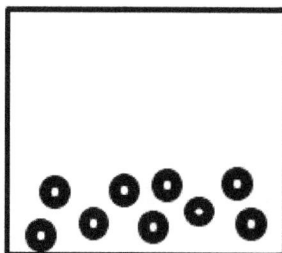

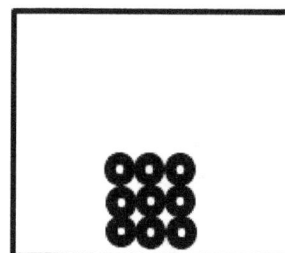

Name_____

Self-Test 1.5 Scientific Measurement and Significant Figures

1. Match the metric units associated with each of the symbols:
    a) ____ 0.001 m   A) μm

    b) ____ 10^{-2} m   B) cm

    c) ____ 1 x 10^{-9} m   C) mm

    d) ____ 1/1,000,000 m   D) nm

2. Make the following conversions using the unit cancellation method. **Show your work and label all calculations.**

    a) 47 L = ? mL   b) 44.6 kg = ? g

    c) 3.77 cm = ? nm   d) 27 mm = ? cm

    e) 300 cm^3 = ? L   f) 1500 μL = ? mL
    (hint: 1 cm^3 = 1 mL)   (hint: 1000 μL = 1 mL)

3. Give the number of significant figures in each of the following measurements:
    a) 275 °C _____
    b) 0.0256 L _____
    c) 2002 g _____
    d) 2.50 x 10^2 cm_____
    e) 5600 J_____

4. Express 1000 mL using two significant figures._____

5. Solve the following using significant figure rules:
    a) <u>32.54 g</u> =   b) 56.795 g   c) 534.89 mL   d) 32.115 m
       4.5 mL   <u>+ 1.2   g</u>   <u>- 12   mL</u>   1.8011 m
                                                                         179.32   m
                                                                      <u>+   5.66   m</u>

e) 8.451 cm • 13.6 cm • 5.4 cm =   f) 567.90 mm • 4.6 mm =

6.
- a) The following data were obtained. Calculate the average of the trials.
  Trial 1: 4.45 g/mL
  Trial 2: 4.55 g/mL
  Trial 3: 4.50 g/mL

- b) The actual value is 4.52 g/mL. Calculate the percent error.

7. Calculate the density of a cylinder in <u>two ways</u> using the data below.
    SHOW ALL WORK AND LABELS FOR FULL CREDIT. USE CORRECT S.F.

Method 1	Method 2
mass = 60.00 g	mass = 60.00 g
height = 7.0 cm	Volume of water without cylinder = 30.0 mL
diameter = 2.0 cm	Volume of water with cylinder = 53.0 mL
(the volume for a cylinder is $\pi r^2 \cdot h$)	

8. Which method above is more precise and why is it more precise?

Name_____

Review Unit 1

1. Mark the following as element (E) or compound (C).

   a) _____ Tin (Sn)

   b) _____ Glucose ($C_6H_{12}O_6$)

   c) _____ Hydrogen gas (H_2)

   d) _____ Water (H_2O)

2. Classify each of the following substances in the table below (choose all that apply):

Substance	Pure Substance or Mixture?	Heterogeneous or Homogeneous?
river water		
smoke from a fire in air		
uranium (U)		
air		
potassium chloride (KCl)		
garden soil		

3. Mark each of the following as either a physical (P) or a chemical (C) change:

   a) _____ ice cream melting   b) _____ gasoline burning

   c) _____ carrots being chopped   d) _____ soda pop going flat

   e) _____ iron rusting   f) _____ food waste decomposing

4. In the blank next to the statement about chlorine gas (Cl_2), tell whether a physical (P) or chemical (C) property is indicated:

   a) _____ Chlorine is a pale, green gas.

   b) _____ When chlorine is combined with hydrogen, a colorless, gaseous compound is formed.

   c) _____ Chlorine condenses to a liquid at 239K.

   d) _____ When magnesium is exposed to chlorine, magnesium becomes coated with a white, brittle solid.

   e) _____ Chlorine has a density of 3.214 g/mL.

CALCULATIONS: SHOW ALL WORK, UNIT LABELS, AND USE CORRECT SIG FIGS.

5. Calculate the density of a cylinder with a diameter of 1.92 cm, a height of 3.35 cm, and a mass of 64.590 g.

6. A graduated cylinder is weighed dry, filled to the 25.0 mL mark with a liquid and weighed again. From the data below, calculate the density of the liquid.
   Mass of dry graduated cylinder = 18.43 g
   Mass of graduated cylinder with the liquid = 38.65 g
   Volume of liquid in the graduated cylinder = 25.0 mL

7. If a student finds the density of a sample of lead to be 10.65 g/mL in the lab, what is her percent error? (Densities of elements can be found on the periodic table.)

8. How many significant figures are in the following measurements?

a) 0.023 g _____   b) 8.67490 m _____   c) 12.26 cm³ _____

d) 100 yds _____   e) 0.00040 mL _____   f) 35.007 L _____

9. If a string is 5.10 **dm** long, calculate how much string must be cut off to obtain a string that is 50.1 **cm** long. **SHOW ALL WORK.**

10. What type of change is represented below? _____

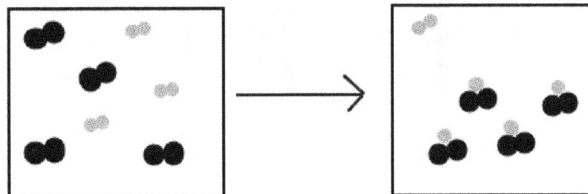

11. Sample 1 of pure aluminum weighs 5.0 grams while sample 2 of pure aluminum weighs 50.0 grams. For the properties listed below, mark A if sample 1 is less than sample 2, mark B if sample 1 is greater than sample 2, and mark C if sample 1 is the same as sample 2.

a) _____ volume

b) _____ melting point

c) _____ density

d) _____ mass

12. Make the following conversions: **SHOW ALL WORK**

a) 100.50 mL = _____L

b) 1 cm^3 = _____mL

c) 650 nm = _____cm

d) 4380 μg = _____mg

13. Match the boxes with the descriptions given below:

a)
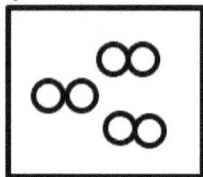

b)
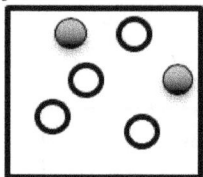

c)
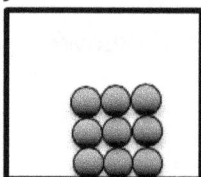

d)
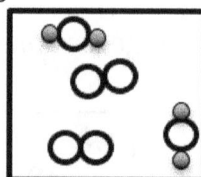

e)
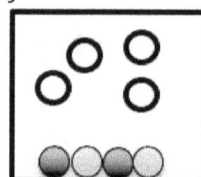

_____ Contains a heterogenous mixture of atoms

_____ Contains a solid only

_____ Contains a heterogenous mixture of molecules

_____ Contains substances in different phases

_____ Contains a pure substance

Unit 2
Atomic Theory and the Structure of Matter

Did you know that?

Some elements on the periodic table are man-made.

Multiple forces hold an atom together.

Archaeologists use isotopes to date artifacts.

Iron filings may be healthier than iron supplements.

Positively charged particles in hair conditioners reduce static electricity between hair strands.

The sky is blue because the wavelength of blue light is scattered by the atmosphere.

Microwaved foods are less dangerous than oven baked foods.

There is no pigment color where there is no light.

Pigment dyes are made from the transition metal compounds.

Wearing white on a hot summer day can keep you cooler than wearing darker colors.

Household incandescent bulbs are the least efficient light bulbs.

X-rays can be used to authenticate a painting.

Several minerals will fluoresce when exposed to UV light.

Chemical reactions in certain marine animals can cause luminescence.

UV lights are often used in forensics.

Objectives for Unit 2

- I can use the periodic table to determine the number of protons, electrons, neutrons, atomic number, and atomic mass for a neutral element.

- I can calculate the average atomic mass for an element given the percent abundance of each isotope of an element.

- I can describe or draw the current model of the atom and note the location and charge of each of the subatomic particles that make up an atom.

- I can determine the charge on an ion of an element when given the number of electrons.

- I can describe the mathematical relationships between wavelength, frequency, and energy in electromagnetic waves and can use mathematical formulas to calculate values for wavelength, frequency, and energy.

- I can use the terms ground state, excited state, quantum, and photon to explain how an electron in a hydrogen atom behaves during an electron transition.

- I can write the electron configuration and the noble gas configuration for an element in the ground state and for an ion of an element.

- I can use Coulomb's Law to describe periodic trends such as atomic radius, ionization energy, and electronegativity.

- I can explain differences in reactivity of elements by using concepts such as ionization energy, atomic radius, and electronegativity.

Name_____

Self-Test 2.1 Atoms, Ions, and Isotopes

1. Use the periodic table to complete the following table. Include the symbol of the isotope or ion, number of protons, neutrons, electrons, and the atomic number for each.

Isotope	Number of p+	Number of e-	Mass number	Number of n⁰	Atomic number
$^{12}_{6}C$					
	5		10		
	5		11		
			9		4
$^{37}_{17}Cl$					
Na⁺				12	
2+					56
2-		10		8	

2. Fill in the following table regarding the subatomic particles.

Particle	Symbol	Charge	Location (inside the nucleus or outside)	Actual Mass	Order from smallest (1) to largest (3)
Electron					
Proton					
Neutron					

3. Which subatomic particle determines the identity of an atom?_____

4. Which subatomic particle determines the behavior of the atom?_____

5. Which subatomic particles determine the mass of the atom?_____

6. For each of the following ions, determine if the atom gained or lost electron(s) by writing a G for gained and an L for lost.

    a) _____ F^-

    b) _____ K^+

    c) _____ Ca^{2+}

    d) _____ S^{2-}

Self-Test 2.2 Average Atomic Mass

1. What is an isotope?

2. Using the interactive periodic table (www.ptable.com), list the selected isotopes for oxygen and the percent abundance of each.

3. Fill in the following table for oxygen-16 and oxygen-18.

Subatomic Particle	oxygen-16	oxygen-18
number of p^+		
number of e^-		
number of n^0		

4. The following graph shows the percent abundance of each isotope of magnesium.

Graph 2.2 Percent Abundance of Magnesium

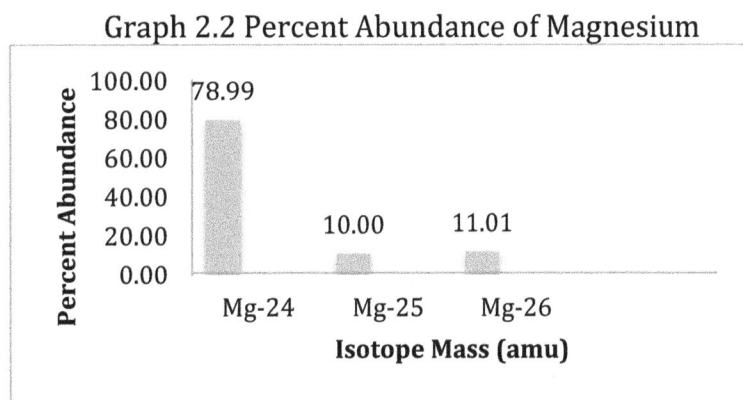

Fill in the following operation to calculate the average atomic mass of magnesium. Use the percent abundance to determine the number of significant digits in the answer. Include units.

(0.7899)(24 amu) + (   )(25 amu) + (0.1101)(26 amu) = (   )

5. The atomic masses listed on the periodic table are averages of the isotopes for each element. Given the following isotopic composition:

    $^{28}Si = 92.23\%$ $^{29}Si = 4.68\%$ $^{30}Si = 3.09\%$

    a. Predict which isotope the average atomic mass will be closest to.

    _____

    b. Calculate the average atomic mass for silicon. Show all calculations and make sure to use proper sig figs. (Remember to convert the percentages into correct decimal values by dividing by 100!)

6. Calculate the average atomic mass of boron (B). B-10 has a percent abundance of 19.9% and B-11 has a percent abundance of 80.1%. **SHOW ALL CALCULATIONS BELOW AND USE PROPER SIG FIGS.**

7. A mass spectrometer separates atoms based on mass. After passing through a beam of high-speed electrons, the atoms become positively charged ions. As these ions are accelerated in a magnetic field, which of the following is/are true? Refer to figure 5-1 in the text reading on Atomic Mass.

    a) The ion with the largest mass will be deflected by the magnetic field the most because the larger mass is more affected by the applied magnetic field.

    b) The ion with the smallest mass will be deflected by the magnetic field the most because the lower mass has less resistance to the pull of the field.

    c) The atom with the smallest mass will move the fastest and will be deflected the least because it can escape the pull of the magnetic field.

Name_____

Self-Test 2.3 The Electromagnetic Spectrum

1. Light is a form of energy found on the electromagnetic spectrum. All electromagnetic radiation travels in waves. Label the diagram of the wave below with the following: **high energy, low energy, long wavelength, short wavelength, high frequency, low frequency**.

   _____   _____

   _____   _____

   _____   _____

2. What distinguishes one part of the EM spectrum, such as microwaves, from another part, such as UV light?

3. Mark each relationship as either direct (D) or indirect (I).

   a. _____ The relationship between wavelength and frequency.

   b. _____ The relationship between wavelength and energy.

   c. _____ The relationship between frequency and energy.

4. Order the following types of electromagnetic waves from highest energy to lowest energy: visible light, X-rays, infrared waves, microwaves, UV rays.

   highest   _____

              _____

              _____

              _____

   lowest   _____

5. Microwaves, yellow visible light, and UVB radiation differ in their wavelength, frequency, and energy.

Type of Radiation	Wavelength
Microwave	0.50 m
Yellow Light	580 nm
UVB	290 nm

Useful equations and constants

$c = \lambda v$
λ = wavelength
v = frequency
$c = 3.00 \times 10^8$ m/s (speed of light)
$E = hv$
$h = 6.626 \times 10^{-34}$ J•s

a) Determine the **frequency** (v) associated with each type of radiation. *The wavelength must be in meters. SHOW ALL WORK AND UNITS. INCLUDE SIG FIGS.

Microwave:

Yellow Light:

UVB:

b) Calculate the **energy** (E) in each type of radiation. SHOW ALL WORK AND UNITS.

Microwave:

Yellow Light:

UVB:

6. Using your answers from #5b above, explain why UV radiation can cause skin cancer but light from an incandescent bulb does not.

7. How is microwaving food different than cooking with heat (baking or grilling)? Which method could be more damaging to the nutrients in the food?

8. T/F Violet light travels faster than red light.

Name_____

Self-Test 2.4 The Bohr Model of the Atom

1. Ernest Rutherford's model of the atom included a central concentration of charge surrounded by a cloud of orbiting electrons. Niels Bohr proposed a more detailed model of the atom.
   a) What does the Bohr model of the atom look like (draw and describe).

Picture	Description

2. Fill in the appropriate term to complete the sentence.
   a) A specific amount of energy required to move an electron to a higher energy level is called a _____.

   b) Electrons that are _____ to the nucleus have lower energy.

   c) The term used to describe the lowest energy level of an electron is the _____ state.

   d) An electron is in the _____ state when it absorbs energy of a specific quantum.

   e) When an electron returns to the lowest energy level from a higher energy level a _____ of energy is emitted.

   f) The_____ _____ _____ shows the visible colors of light that are given off when energized electrons in an atom return to their lowest energy state.

3. The emission spectrum of hydrogen consists of four visible colored lines. What do the lines represent?

4. If you are trying to find the location of your phone in a dark room you would likely turn on a light to find it. Why won't this method work to determine the exact location of an electron in an atom?

Use the following information for the energies associated with the hydrogen atom:

Principle Energy Level (n)	Energy in Joules (J)
Infinity	0
6	-6.05 x 10^{-20}
5	-8.72 x 10^{-20}
4	-1.36 x 10^{-19}
3	-2.42 x 10^{-19}
2	-5.45 x 10^{-19}
1	-2.18 x 10^{-18}

The following information may be useful in solving these wave problems.

Constants:

$c = 3.00 \times 10^8$ m/s

$h = 6.626 \times 10^{-34}$ J•sec

Equations:

$$\lambda = \frac{hc}{\Delta E}$$

*λ must be in meters

5. Calculate the change in energy when an electron drops from n=5 to n=3 in hydrogen.

6. Calculate the wavelength of energy given off from the electron transition in problem 5.

7. Calculate the wavelength of light when an electron drops from n=5 to n=2.

8. Convert this wavelength to nanometers and determine the color of light given off in this transition. (Violet = 400 nm, Red = 700 nm)

Name_____

Self-Test 2.5 Electron Configurations in the Ground State

1. The following questions refer to the electron configuration: $1s^2 2s^2 2p^6 3s^2 3p^6 4s^1$

    a) What element is this?_____

    b) What is the total number of protons in the nucleus of this element?____

    c) How many electrons are in the outer shell of this element?_____

    d) What is the atomic mass of this element?_____

2. Write the ground state electron configuration for the following elements and determine the number of valence electrons and total electrons in each.

Element	Ground State e- Configuration	# of Valence Electrons	Total # of electrons
carbon			
magnesium			
bromine			
oxygen			
potassium			
iron			
neon			
nitrogen			
lead			

3. Determine the ground state noble gas configuration for the following elements:

    a) phosphorus_____

    b) tellurium_____

    c) strontium_____

Electron Configurations of Ions

4. How many electrons do all Noble Gases have in the <u>outer shell</u>?_____

5. Look at the ground state configuration for the following atoms and answer the questions in the table.

Ground State Configuration	$1s^22s^22p^63s^23p^1$	$1s^22s^22p^63s^23p^5$	$1s^22s^22p^63s^23p^64s^2$
What element is this?			
How many p+ are in the nucleus of this element?			
How many valence e- are in this element?			
Will this element lose or gain e- to achieve a stable configuration?			
What ion will most likely be produced from this element? (Write the symbol, + or -, and the numerical value of the charge.)			
Write the ground state electron configuration for the ion produced.			
Does the ion have a configuration like a Noble Gas? (Y/N)			

6. When an atom of an element becomes an ion, does it turn into a Noble Gas? Explain your answer by addressing the protons in the nucleus.

7. Does the atomic mass of an atom of an element change when it becomes an ion? Why or why not?

8. In order for magnesium to become an ion (Mg^{2+}), the magnesium must:
   a) lose two electrons

   b) lose two protons

   c) gain one neutron

   d) gain one electron

Name_____

Self-Test 2.6 Electrostatic Forces and Periodic Trends

To fully explain periodic trends, it is important to understand electrostatic forces and concepts related to Coulomb's Law, $F_e = \dfrac{kq_1q_2}{d^2}$. Use Coulomb's Law to answer the following questions:

Coulomb's Law

1. There is a(n) _____ (direct/indirect) relationship between the amount of charge and the force of attraction or repulsion.

2. There is a(n) _____ (direct/indirect) relationship between the distance between two charged objects and force of attraction or repulsion between the objects.

3. Like charges _____, unlike charges _____.

4. The force between the nucleus and an electron in an atom _____ as the distance between them increases. This means that it requires less energy to remove an electron from an orbital that is _____ (closer to/farther from) the nucleus. Atoms of elements with more protons and electrons will have a _____ (stronger/weaker) force of attraction between the electrons and protons.

Periodic Trends: Atomic Radius

Atomic radius is the distance from the nucleus to the <u>probable</u> location of the outermost electron.

5. Find atomic radius on the periodic table and describe the general trend in atomic radius from Group I to Group VII.

   a) Describe the trend in atomic radius as you move down Group I.

   b) Draw arrows on the periodic table below to show the direction that the atomic radius <u>increases</u> in a period and a group.

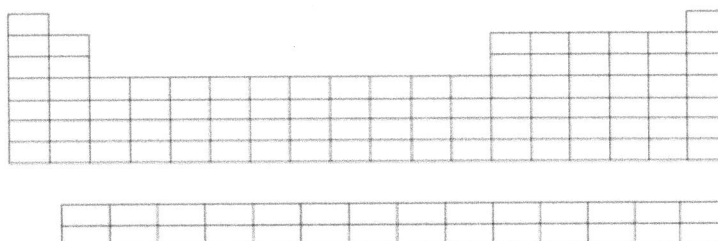

6. Using Coulomb's Law, explain why the atomic radius decreases as you move from left to right across a period.

7. Refer to the diagram below of Na and K to explain why atomic radius increases as you move down a group.

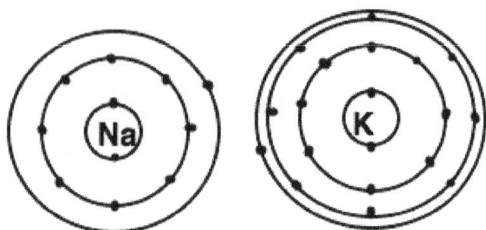

Periodic Trends: Ionic Radius

8. When an atom becomes an ion, it either loses or gains one or more electrons, and the radius changes size. Look at the diagram of sodium, which loses one electron to become Na^+.

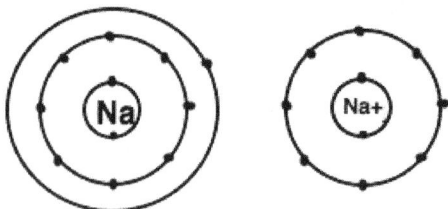

Describe what happens to the radius.

9. Now, look at the diagram of fluorine (F), which gains an electron to become F^-.

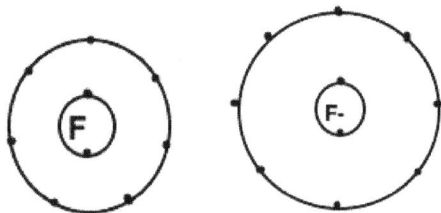

Describe what happens the radius.

10. When an atom becomes an ion, the new radius is called the <u>ionic radius</u>.

    a) Compare atomic radius and ionic radius of several elements that form positive ions. What generalization can be made about what happens to the size of the radius of an atom that forms a positive ion (cation)?

    b) Why does this occur?

    c) Compare atomic radius and ionic radius of several elements that form negative ions. What generalization can be made about what happens to the size of the atomic radius of an atom that forms a negative ion (anion)?

    d) Knowing how electrons behave around other electrons, describe why this happens.

Periodic Trends: Ionization Energy

<u>Ionization energy</u> is the amount of energy that is required to <u>remove</u> an electron from an atom.

11. Find 'First Ionization Potential' on the periodic table and describe the general trend as you move from left to right across a period.

    a) Using Coulomb's Law, explain why this trend occurs by relating it to what happens to the atomic radius as you move from left to right across a period.

    b) Describe the general trend in ionization energy as you move down a group on the periodic table.

    c) Explain this trend by relating it to what happens to the atomic radius as you move down a group.

    d) What element has the highest First Ionization Potential? Why?

    e) Is the relationship between atomic radius and ionization energy direct or indirect?

Periodic Trends: Electronegativity

Electronegativity is a measure of an element's ability to attract an electron in order to become stable.

12. Find electronegativity on the periodic table and describe the general trend as you move from left to right across a period.

   a) Moving from left to right increases the number of protons in the nucleus; how does this factor affect the electronegativity?

   b) Describe the general trend in electronegativity as you move down a period.

   c) While the number of protons in the nucleus increases as you move down a group, consider what happens to the atomic radius; how does the atomic radius affect the electronegativity of an element?

   d) What is the most electronegative element? Why?

Elements that are unstable are reactive. The most reactive elements tend to be in Group I and in Group VII.

13. Considering what you know about atomic radius and ionization energy, which element in Group I do you think would be the most reactive? Why?

14. Considering what you know about atomic radius and electronegativity, which element in Group VII do you think would be the most reactive? Why?

15. Can a general trend for reactivity be stated or does it depend on the group that the element is in?

Name_____

Review Unit 2

1. Calculate the average atomic mass for argon given the following isotopes. SHOW ALL WORK.

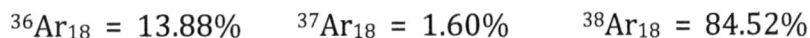

$$^{36}Ar_{18} = 13.88\% \qquad ^{37}Ar_{18} = 1.60\% \qquad ^{38}Ar_{18} = 84.52\%$$

2. Complete the table:

Symbol	Number of p+	Number of e-	Number or n^0	Mass number
$^{16}O_8$				
$^{27}Al_{13}$				
	12		12	
		18	18	
$^{40}Ca_{20}^{2+}$				

3. A sample of element X is charged and analyzed in a mass spectrometer. Which isotope of element X will be deflected the least by a magnetic field? Circle the correct answer.

$$^{121}X \qquad\qquad ^{123}X \qquad\qquad ^{125}X$$

a) Explain how you know this.

4. Mark T if the statement is true about a neutral atom with an atomic number of 33 and a mass of 75 and F if it is false.

    a. _____ It has 33 protons.

    b. _____ It has 42 electrons.

    c. _____ It has 33 neutrons.

    d. _____ It has 75 neutrons.

    e. _____ It has 75 neutrons and electrons.

5. Circle the following pairs that are isotopes of each other.

    a. $^{16}O_8$ and $^{16}N_7$

    b. $^{4}He_2$ and $^{4}He_3$

    c. $^{14}N_7$ and $^{15}N_7$

    d. $^{1}H_1$ and $^{4}He_2$

6. T/F When an atom loses an electron it becomes negatively charged and is called an anion.

7. Fill in the table regarding the following ions:

Symbol	# of p^+	# of n^0	# of e^-	Oxidation # or charge
$^{16}O_8{}^{2-}$				
$^{27}Al_{13}{}^{3+}$				
	12	12		2^+
		19	18	1^-

8. For the ground state, outer shell electron configuration for a neutral atom:
$1s^2 2s^2 2p^6 3s^2 3p^6 4s^2 3d^{10} 4p^4$

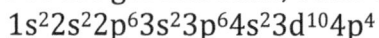

a) How many valence electrons are in this atom?_____

b) What atom is this?_____

c) What ion will this atom most likely form?_____

d) Write the electron configuration for the ion formed._____

e) What is the atomic number of this element?_____

9. Calculate the frequency of visible light that has a wavelength of 620nm. **SHOW YOUR WORK AND INCLUDE LABELS.**

10. Calculate the wavelength of light given off when an electron makes the transition from n=4 to n=2 in the hydrogen atom. Use the energy values from the table on ST 2.4.

11. Fill in each of the following periodic trends using labeled arrows to show an increase in the trend: atomic radius, first ionization, and electronegativity.

Unit 3
Chemical Bonds

Did you know that?

Atoms bond with other atoms to become more stable.

The element Francium is one of the more unstable elements and is rarely found in nature.

Chlorine is a deadly, poisonous gas that is harmless when chemically bonded to sodium.

An atom achieves stability when is has eight valence electrons.

While there are only 118 elements known to man, the number of compounds is uncountable.

Molecules have a three-dimensional structure that influences their interactions with other molecules.

Water molecules have a special shape and arrangement of electrons that make the molecules stick together and enable water striders to "walk on water".

Objectives for Unit 3

- I can determine if a substance contains an ionic bond, nonpolar covalent bond, or polar covalent bond based on the chemical formula for the substance.

- I can determine the relative strength of the interatomic bond in different substances when given the chemical formula and bond breaking energies or lattice energies.

- I can draw the Lewis structure and standard notation for covalent molecules and common polyatomic ions.

- I can write the chemical name for covalent compounds, simple organic molecules, ionic compounds, and acids when given the chemical formula.

- I can write the chemical formula for covalent compounds, simple organic molecules, ionic compounds, and acids when given the chemical name.

-Julia Mixon

Name_____

Self-Test 3.1 Types of Chemical Bonds

1. Distinguish the following types of elements on the periodic table below:
   a) metals   ▨▨▨
   b) nonmetals (don't forget to include hydrogen)   ▮▮▮

2. Covalent compounds are made up of
   a)metals only   b)nonmetals only   c)both metals and nonmetals

3. Describe how the atoms in a covalent bond achieve a noble gas configuration (the most stable configuration).

4. Name the two nonmetal elements that achieve stability with two valence electrons:
   a)_____   b)_____

5. Ionic compounds are made up of
   a)metals only   b)nonmetals only   c)both metals and nonmetals

6. Describe how the atoms in an ionic bond achieve a noble gas configuration.

Bond Polarity

The degree that atoms share or transfer electrons determines the character of the bond within the compound. Atoms in ionic compounds transfer electrons resulting in oppositely charged ions that are attracted to each other. By nature, an ionic compound has two "poles" or oppositely charged sides. In a covalent compound or molecule, electrons can be shared equally or unequally depending on the electronegativity of the atoms in the compound. When atoms share electrons equally, the bond is a **nonpolar covalent bond**. This occurs in molecules containing two of the same nonmetal atom. Any combination of two or more different nonmetal atoms results in unequal sharing of electrons, or a **polar covalent bond**.

The degree of polarity and type of bond can also be determined using <u>electronegativity differences</u>.
- If the difference is greater than or equal to 1.7 and the molecule contains a metal and a nonmetal, the bond is **ionic**.
- If the difference is greater than 0 but less than 1.7, the bond is **polar covalent**.
- If the difference is 0, the bond is **nonpolar covalent**.

7. Fill in the following table. Calculate the electronegativity difference, the type of interatomic bond (ionic, polar covalent, nonpolar covalent), then rank the strength (number from 1 to 10, 1 being the strongest) of the bond based on the energy required to break the bond

Molecule or Compound	Bond breaking energy (kJ/mol)	Electronegativity Difference	Type of Bond	Relative Strength
H_2	436			
KF	821			
HF	569			
HBr	370			
Cl_2	330			
NaCl	787			
LiF	1036			
F_2	160			
HI	295			
KCl	715			

8. Given the relative bond strengths in problem 1, which type of bond is relatively strongest?
   a) covalent
   b) polar covalent
   c) ionic

9. Which compound has a stronger ionic bond?
   a) NaF
   b) NaI
   c) NaBr
   d) NaCl

10. Explain your answer to #2 above using Coulomb's Law.

Self-Test 3.2 Modeling Covalent Compounds

Lewis Dot Structures

1. Use the periodic table to complete the following:

Element	Group Number	# of Valence Electrons	Lewis Dot Structure of the Element
C			
S			
N			
Cl			

2. Fill in the table. The atom with the lowest electronegativity is the central atom. Hydrogen and the halogen group elements are terminal atoms.

Formula	Lewis dot structure for **each** atom in the molecule. Make sure the correct # of valence electrons are shown.	Total number of valence electrons available. Count the electrons on each atom and add them.	Lewis dot structure for the molecule	Standard notation for the molecule. Use lines to represent shared pairs and dots for unshared pairs of e-.
H_2	H• H•	1+1=2	H:H	H-H
HCl				
NH_3				

Formula	Lewis dot structure for **each** atom in the molecule.	Total number of valence electrons available.	Lewis dot structure for the molecule	Standard notation for the molecule.
CH_4				
$CHCl_3$				
H_2O				
CO_2				
CO				
OH^-				
NH_4^+				
CO_3^{2-}				
PO_4^{3-}				

Name_____

Self-Test 3.3 Naming Covalent Compounds

1. Writing formulas for inorganic covalent compounds uses prefixes to tell the number of atoms of each element that make up the compound. Write the number that corresponds to the following prefixes.

____tetra   ____hexa   ____di   ____hepta

____deca   ____tri   ____mono   ____penta

2. Name or write formulas for the following inorganic covalent compounds.

Name of Compound	Use	Molecular Formula
	photochemical smog	NO_2
silicon dioxide	amethyst	
sulfur dioxide	London smog	
	auto exhaust	NO
dinitrogen monoxide	laughing gas	
carbon tetrachloride	once used for cleaning	
	brown byproduct of smog	N_2O_6
	fertilizer	P_4O_6
carbon dioxide	greenhouse gas	

Organic Hydrocarbons:

3. Organic hydrocarbons are a special class of covalent compound that consist of varying numbers of only carbon and hydrogen atoms. Different prefixes are used to note the number of <u>carbon atoms</u> in the molecule. Fill in the number of carbon atoms associated with each prefix:

_____eth-   _____ oct-   _____ but-   _____ hept-

4. Name the following hydrocarbons or write the formula (for alkanes, the simplest hydrocarbon, the number of hydrogen atoms is always two more than twice the number of carbon atoms (n)):

Alkane Name	Molecular Formula (C_n, $H_{(2n+2)}$)
	C_3H_8
decane	
methane	
	C_6H_{14}
pentane	

5. When a double or triple bond exists between the carbon atoms in a hydrocarbon, the hydrocarbons are called alkenes and alkynes respectively. The suffix of the name changes from –ane to –ene (for a double bond) and –yne (for a triple bond). The number of hydrogen atoms decreases due to the multiple bond (2n for alkenes and 2n-2 for alkynes). The location of the double or triple bond is also noted. See the example in the table below:

Hydrocarbon name	Molecular formula ($C_n,H_{(2n)}$ or $C_n,H_{(2n-2)}$)	Lewis Structure
| 2-pent<u>ene</u><br>(The –ene signals that there is a double bond, the 2 means the double bond is between the second and third carbon.) | $C_5H_{10}$ | ```
H   H H H
|   | | |
H-C-C=C-C-C-H
| |   | |
H H   H H
``` |
ethyne		
	1-butyne	
3-hexene		
```

Name_____

Self-Test 3.4 Ionic Bonding and Ionic Compounds

1. Use the periodic table to predict the charge on an atom of each of the following elements when an ion is formed, if it is a transition metal, note the two most common charges:

Element	Ion Formed
oxygen	O^{2-}
potassium	
iodine	
calcium	
nitrogen	
lithium	
aluminum	
iron	
nickel	Ni^{2+}, Ni^{3+}
lead	
zinc	
sulfur	

2. Fill in the table.

Binary Ionic Compounds

Cation (metal)	Anion (nonmetal)	Formula (must be balanced)	Name	Use
			cobalt (II) chloride	foam stablizer
Ca^{2+}	Cl^-			Ice Bite
		SnF_2		fluoristan
			tin (IV) fluoride	in toothpaste
Ag^+	S^{2-}			tarnish
		PbO_2		electrode

3. Fill in the table.

Ionic Compounds with Polyatomic Ions

Cation	Anion	Formula (must be balanced)	Name	Use
		Na_3PO_4		grease cleaner
Fe^{2+}	SO_4^{2-}			iron pills, ink
		NH_4NO_3		fertilizer
Na^+	SO_4^{2-}			washing soda

4. Fill in the table.

Hydrated Ionic Compounds

Cation	Anion	Formula (must be balanced)	Name
Ca^{2+}	SO_4^{2-}	$CaSO_4 \cdot 2H_2O$	
		$CuSO_4 \cdot 5H_2O$	
			barium hydroxide octahydrate

5. Describe how it can be determined if a compound contains an ionic or a covalent bond based on the formula.

Name_____

Self-Test 3.5 Naming Acids and Bases

An inorganic acid is a type of covalent molecular compound that produces hydrogen ions (H^+) when dissolved in water. Inorganic acids typically contain one or more hydrogen atoms combined with a nonmetal ion or polyatomic ion. Formulas for acids must be balanced with the correct number of hydrogen ions (H^+) to balance the negatively charged nonmetal ion or polyatomic ion.

1. Fill in the following table with the name of the acid or the molecular formula for the acid.

Molecular Formula	Name of Acid	Use
	phosphoric acid	fertilizer production
H_2CO_3		carbonated beverages
	sulfuric acid	battery acid
H_2SO_3		acid rain
	hydrofluoric acid	etches glass
HCl		stomach acid
	nitric acid	fertilizer production

Bases are ionic compounds that produce hydroxide ions (OH^-) when dissolved in water. They are named the same as ionic compounds and will usually end in hydroxide since most bases contain hydroxide ions.

2. Fill in the following table with the name of the base or the molecular formula for the base.

Ionic Formula	Name of Base	Use
$Ca(OH)_2$		plaster
	sodium hydroxide	drain cleaner
NH_4OH		leavening agent
	potassium hydroxide	used in soap
	ammonia (look this up)	cleaner
	aluminum hydroxide	deodorant

Name_____

Review Unit 3

1. Determine the type of bond formed in the following compounds then name the compound or write the formula.

Compound	Type of Bond	Name of Compound
ZnO		
CCl$_4$		
SO$_3$		
Ca$_3$(PO$_4$)$_2$		
		dinitrogen pentoxide
		ammonium fluoride
		titanium (IV) oxide
		hydrogen chloride

2. Fill in the following table with the name or formula for the compounds.

Chemical Formula	Name of Compound
	iron (II) nitrate
	tin (IV) chloride
H$_2$CO$_3$	
	ammonium sulfate
	sulfur trioxide
	iron (III) sulfide
	hydrobromic acid
Na$_2$CO$_3 \cdot$10H$_2$O	
Fe$_2$O$_3$	
Cu$_2$S	
NO$_2$	
	aluminum hydroxide
	magnesium hydroxide heptahydrate
	aluminum sulfate octahydrate
P$_2$O$_5$	
C$_6$H$_{14}$	
	ethane

3. Draw the Lewis dot structure and the structural formula for the following molecules to determine if they contain a double or a triple bond.

Substance	Lewis dot	Structural formula
C_2H_4		
C_2H_3Cl		
CO		
SO		
NH_3		
H_2S		
HI		

Unit 4
The Mole and Stoichiometry

Did you know that?

Every breath you take contains molecules that people such as Leonardo da Vinci and Shakespeare once breathed.

It takes billions, and billions, and billions of water molecules to fill up your bathtub.

The mole is a word to describe multiple things, including a unit of measurement in chemistry.

One teaspoon of table sugar (sucrose) contains 7.4×10^{21} molecules of sugar, and it takes 8.9×10^{22} molecules of oxygen to metabolize all that sugar.

Bakers are really specialized chemists; they combine ingredients in exact proportions to create a desired food product.

Objectives for Unit 4

- I can convert quantities of substances from grams to moles to molecules and/or atoms as well the reverse.

- I can calculate the molecular mass of a substance when given the chemical formula.

- I can determine the percent composition of an element in a compound and the number of grams of an element that is present in a given amount of a compound.

- I can determine the empirical and molecular formulas for a compound when given the percent composition of each element and the molar mass of the compound.

- I can write the chemical formulas and chemical equations for a reaction when given the chemical names of the substances in the reaction.

- I can balance chemical equations to represent conservation of mass.

- I can determine the limiting reactant, excess reactant, and theoretical yield of a product when given initial quantities of reactants present before a chemical reaction. When experimental yield is also given, I can determine the percent yield and percent error.

- I can determine the amount of each reactant needed to make a given amount of a product in a chemical reaction.

- I can use molecular models to determine the limiting reactant, excess reactant, and amount of product formed in a chemical reaction.

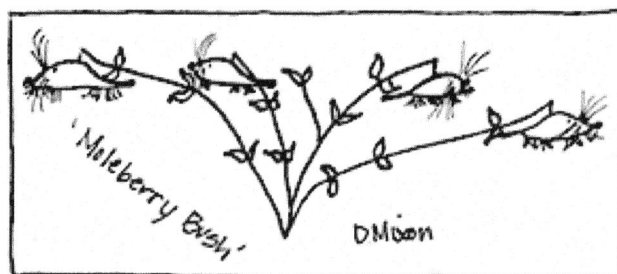

Moleberry Bush, DMoon

Name_____

Self-Test 4.1 The Mole

Example Problem: Calculate the number of atoms in 3.91 g of aluminum.
   Steps:
   1. Convert grams to moles by dividing by the atomic mass of aluminum.
   2. Convert moles to atoms by multiplying by Avogadro's number.

$$3.91 \text{ g Al} \times \frac{1 \text{ mol}}{26.98 \text{ g}} \times \frac{6.022 \times 10^{23} \text{atoms}}{1 \text{ mol}} = \boxed{8.73 \times 10^{22} \text{ atoms Al}}$$

          3 s.f.   units   substance label

SHOW ALL WORK, INCLUDE UNITS AND LABELS, AND USE CORRECT SIG FIGS

1. Calculate the number of atoms in 7.7 moles of gold.

2. Calculate the number of moles of calcium that are needed daily if 1300 mg are required.

3. Calculate the number of moles of copper present in 9.15×10^{23} atoms of copper.

4. Calculate the number of atoms in 12.80 g of sodium.

5. Calculate the mass of 3.29×10^{23} atoms of nickel.

6. Explain why the mole is used as a unit of measurement in atoms. Use complete sentences.

7. In a particular lab, you are required to measure out 0.600 moles of carbon. Explain how you will do this by showing the calculations below.

Name_____

Self-Test 4.2 Molecular Mass

Example Problems

Find the molecular mass of calcium carbonate, $CaCO_3$, the substance that chalk is composed of.

I atom of Ca = 40.08 g/mol
+ I atom of C   = 12.01 g/mol   Add the individual atomic masses together to find the
3 atoms of O
(3 x 16.00) = 48.00 g/mol
            100.09 g/mol $CaCO_3$ Note the answer has correct sig. figs., units, and label.

Calculate the number of moles in 35.5 grams table sugar ($C_{12}H_{22}O_{11}$).

Step 1: Find the molecular mass of $C_{12}H_{22}O_{11}$.	Step 2: Convert grams to moles by dividing by molecular mass.
12 atoms C (12 x 12.01g/mol) = 144.12g/mol +22 atoms H (22 x 1.01 g/mol) = 22.22g/mol 11 atoms O (11 x 16.00 g/mol)=176.00g/mol 342.34 g/mol	35.5 g $C_{12}H_{22}O_{11}$ x $\dfrac{1\ mol}{342.24\ g}$ = 0.104 mol $C_{12}H_{22}O_{11}$ Note the <u>courtesy zero</u> before the decimal, 3 s.f., units, and substance label

SHOW ALL WORK, INCLUDE UNITS AND LABELS, AND USE CORRECT SIG FIGS

1. Calculate the number of moles there would be in each of the following:
   a) 7.5 g of baking soda ($NaHCO_3$).

   b) 57.0 g of antiperspirant ($Al_2(OH)_5Cl \bullet 2H_2O$). Note that $\bullet 2H_2O$ means that you must add the mass of two water molecules, you <u>do not</u> multiply by the mass of two water molecules.

2. Calculate the mass of each of the following:
   a) 10.25 moles of household ammonia (NH_3).

b) 8.35 moles of natural gas (CH_4).

Example Problems

a) Calculate the number of molecules in 3.50 moles of salt (NaCl).

$$3.50 \text{ moles NaCl} \times \frac{6.022 \times 10^{23} \text{ molecules}}{1 \text{ mol}} = 2.11 \times 10^{23} \text{ molecules NaCl}$$

b) How many atoms are in 2.11×10^{23} molecules (3.50 moles) of NaCl?

1 molecule of NaCl contains two atoms (1 Na + 1 Cl). Multiply the # of molecules by 2 atoms.

$$2.11 \times 10^{23} \text{ molecules NaCl} \times \frac{2 \text{ atoms}}{\text{molecule}} = 4.22 \times 10^{23} \text{ atoms in 3.50 moles NaCl}$$

3. Calculate the number of molecules of carbon dioxide (CO_2) in a fire extinguisher that contains 2.15 kg of CO_2. (Note the amount is in kg – convert first.)

4. Calculate the mass of 7.45×10^{23} molecules of table salt (NaCl).

5. Calculate the total number of atoms that make up 100.4 g distilled water (H_2O).

Name_____

More Moles!

The goal of this activity is to determine the number of molecules in common, household substances.

You and your partner will determine how you will do the following:

1. Find the number of molecules in a sugar cube and the total number of atoms of <u>each element</u> (the chemical formula for sucrose is $C_{12}H_{22}O_{11}$).
2. Find the number of molecules of glass (silicon dioxide) in a marble.
3. Find the number of molecules of water in a small ice cube and the total number of atoms of each element.
4. Find the number of molecules of chalk (calcium carbonate) that it takes to write your full name on a chalkboard.
5. Bring Ms. Mixon 5.2×10^{22} molecules of sodium chloride.
6. Determine the number of molecules of water in popcorn kernel (popping the kernel releases the water by turning it into $H_2O_{(g)}$).
7. Determine the percent by mass of water in a popcorn kernel.

Make a table of what you will need to measure here. Label it appropriately and use correct units.

Do your calculations here. SHOW WORK, UNITS, LABELS, AND USE CORRECT SIG FIGS.

Compound Name and Formula	Calculations	Answer
sucrose $C_{12}H_{22}O_{11}$		$C_{12}H_{22}O_{11}$
		C:
		H:
		O:
silicon dioxide _____		
water _____		
		H:
		O:
chalk _____		
sodium chloride _____		
water vapor $H_2O_{(g)}$		
Percent by mass of water in kernal		

Name_____

Self-Test 4.3 Percent Composition

1. Pure gold is too soft to be worn as jewelry so other metals are added to gold to make it more durable. The percent composition of gold in a sample is known as karatage (k). 24 k is 100% gold, 18 k is 75% gold, and 14 k is 58.3 % gold.

    a) A sample of gold jewelry has a mass of 85.5 g. Once refined, 64.12 g of gold is recovered. What is the percent composition of gold in the sample? **SHOW ALL CALCULATIONS**

    b) How many karats was the original jewelry?

2. A mineral sample is known to contain one form of a compound of iron oxide, either Fe_2O_3 or Fe_3O_4. Following a reduction reaction of the iron ore, the iron is purified and massed to determine the original compound. The following table contains the data:

    | Original Mass of Compound (g) | 137.0 |
    | Mass of Filter Paper (g) | 2.3 |
    | Mass of Filter Paper + Iron (g) | 98.2 |

    a) Find the final mass of pure iron. **SHOW ALL CALCULATIONS**

    b) Find the percent composition of iron in the sample.

    c) Determine which form of iron oxide the sample was composed of.

3. What is the mass of water contained in a 50.0 g sample of magnesium sulfate heptahydrate? **SHOW ALL CALCULATIONS**

4. A sample of a hydrate of calcium sulfate was heated in a crucible to remove the water, forming anhydrous calcium sulfate. Refer to the data in the table to determine the percent of water in the hydrate. **SHOW ALL CALCULATIONS**

Mass of crucible (g)	35.45
Mass of crucible + hydrate (g)	45.45
Final Mass of crucible + anhydrate (g)	43.36

Name_____

Self-Test 4.4 Determining Empirical and Molecular Formulas

1. A compound is analyzed and found to contain 69.55% oxygen and 30.45% nitrogen. What is the empirical formula for the compound?

2. The molar mass of the compound in question 1 is 92.02 g/mole. Determine the molecular formula for the compound.

3. The empirical formula of an organic molecule is CH_2O. The molar mass of the substance is 180.18 g/mol. What is the molecular formula for this compound?

4. An unknown hydrocarbon is analyzed and found to contain 82.63% carbon and 17.37% hydrogen. Determine the formula for the compound.

5. Is the unknown an alkane, alkene, or alkyne? _____

6. Name the compound._____

Name_____

Self-Test 4.5 Balancing Chemical Equations

1. Balance the following reactions:

   a) Magnesium reacts with oxygen gas from the air in flashbulbs:

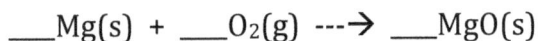

   ___Mg(s) + ___O_2(g) ---→ ___MgO(s)

   b) Milk of magnesia, $Mg(OH)_2$, is an antacid that neutralizes stomach acid (HCl):

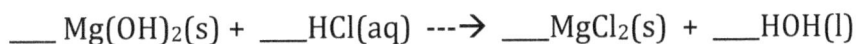

   ___ $Mg(OH)_2$(s) + ___HCl(aq) ---→ ___$MgCl_2$(s) + ___HOH(l)

   c) Acetylene (C_2H_2) is used in an oxy-acetylene torch to produce enough heat to melt steel:

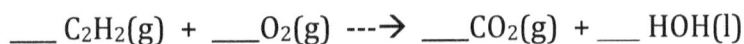

   ___ C_2H_2(g) + ___O_2(g) ---→ ___CO_2(g) + ___ HOH(l)

   d) Fertilizer plants produce ammonia by reacting nitrogen with hydrogen at high temperatures and pressures:

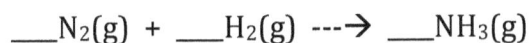

   ___N_2(g) + ___H_2(g) ---→ ___NH_3(g)

   e) Natural gas reacts with oxygen to heat home:

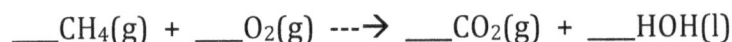

   ___CH_4(g) + ___O_2(g) ---→ ___CO_2(g) + ___HOH(l)

   f) Nitrogen monoxide from automobile exhaust reacts with oxygen to produce toxic smog in Los Angeles:

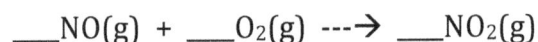

   ___NO(g) + ___O_2(g) ---→ ___NO_2(g)

   g) Iron rusts in moist air to cause millions of dollars of damage each year:

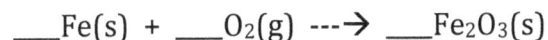

   ___Fe(s) + ___O_2(g) ---→ ___Fe_2O_3(s)

   h) Airbags inflate due to the nitrogen gas produced when sodium azide decomposes:

   ___NaN_3(s) ---→ ___Na(s) + ___N_2(g)

2. Write out and balance the following. Make sure to write your formulas correctly before you balance the equation.

a) Sodium metal reacts with water to make aqueous sodium hydroxide and hydrogen gas (remember that hydrogen gas is diatomic).

b) Aqueous silver nitrate reacts with aqueous copper(II)chloride to make solid silver chloride and aqueous copper(II)nitrate.

c) Aqueous barium nitrate reacts with aqueous sodium phosphate to make aqueous sodium nitrate and solid barium phosphate.

d) Aqueous sodium chloride reacts with aqueous lead(II)nitrate to make solid lead(II)chloride and aqueous sodium nitrate.

Name_____

Cooperative Learning 4.6 Stoichiometric Calculations

1. The balanced coefficients in a chemical equation give the mole to mole ratios in which substances react and are formed.

    Balance: ___$H_2(g)$ + ___$O_2(g)$ ---→ ___$H_2O(l)$

    a) If 2 molecules of H_2 react with 1 molecule of O_2, ___ molecules of H_2O will be formed.

    b) If 4 molecules of H_2 react with 2 molecules of O_2, ___ molecules of H_2O will be formed.

    c) To make 8 molecules of H_2O, ___molecules of H_2 and ___molecules of O_2 are required.

    d) If 2 moles of H_2 react with 1 mole of O_2, ___ moles of H_2O will be formed.

    e) If 4 moles of H_2 react with 2 moles of O_2, ___ moles of H_2O will be formed.

    f) If 4.04 g of H_2 react with 32.00 g of O_2, _____ g of H_2O will be formed. (Remember, mass is conserved).

2. Often, one of the reactants is in excess. This means that there is more of that reactant than can react stoichiometrically with the other reactant. This reactant is called the **excess reactant**. The reactant that is not in excess is called the **limiting reactant**. It is completely used up in the reaction.

    Refer to the following recipe: It takes 6 cups of rice crispies, 2 tablespoons butter, and 40 large marshmallows to make a batch of rice crispy treats. For each scenario below, indicate which ingredient(s) is in excess, which is/are limiting, and how many batches of rice crispy treats can be made. You must use whole numbers only.

Reactants	Excess	Limiting	Batches
18 C rice crispies 8 T butter 80 large marshmallows			
8 C rice crispies 3 T butter 20 large marshmallows			
12 C rice crispies 4 T butter 50 large marshmallows			

3. The first box shows reactants A (squares) and B (circles) before the reaction takes place. The second box shows the result of the reaction.

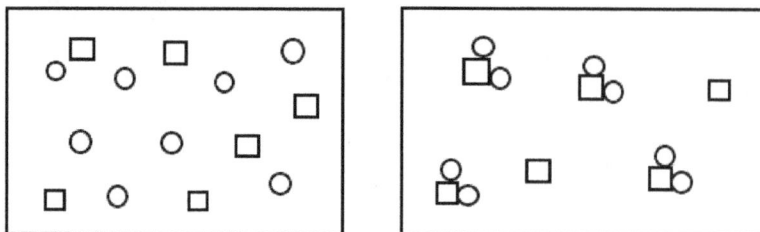

a) Which reactant is the limiting reactant? _____

b) Which reactant is in excess? _____

4. Which of the following shows the balanced reaction that takes place in the boxes shown in problem 3?
    a) $8A(g) + 6B(g) \dashrightarrow 4AB_2(g) + 2A(g)$
    b) $A(g) + 2B(g) \dashrightarrow AB_2(g)$
    c) $A(g) + B_2(g) \dashrightarrow AB_2(g)$
    d) $A(g) + 2B(g) \dashrightarrow AB(g) + 2A$

If you marked B, you were correct. Notice that for every one A, two Bs are required to produce one AB_2. A is in excess, but it is not included in the balanced reaction. Note that both the reactants and the product are in the gas phase, as shown by the distance between particles.

5. The following reaction is used to make iron from iron ore. Balance the equation.

$$\_\_Fe_2O_3(s) + \_\_H_2(g) \dashrightarrow \_\_Fe(s) + \_\_H_2O(g)$$

If the mole amounts of the reactants are known, the mole to mole ratio can be used to predict how many moles of the product will be formed.

To calculate the number of moles of Fe that will be produced when 3.56 moles of Fe_2O_3 react with excess H_2, multiply by the mole to mole ratio. Be sure to show your work and include labels:

$$3.56 \text{ moles } Fe_2O_3 \times \frac{2 \text{ mole Fe}}{1 \text{ mole } Fe_2O_3} = 7.12 \text{ moles Fe}$$

a) How many moles of Fe will be produced from 4.50 moles of H_2 reacting with excess Fe_2O_3?

Name_____

The moles of product produced are converted to grams to calculate **theoretical yield**, which is the amount that should be produced, assuming everything goes perfectly in the laboratory.

b) What is the theoretical yield of Fe (in grams) from 4.50 moles of H_2 reacting with excess Fe_2O_3?

When starting with grams of a reactant (instead of moles), you must first convert the reactant to moles, and then use the mole to mole ratio to find the moles of product produced. Then the moles of product are converted to grams to calculate the theoretical yield.

c) What is the theoretical yield of Fe produced from the reaction of 80.0 grams of Fe_2O_3 with excess H_2?

Percent yield is calculated from theoretical yield and **experimental yield**. Experimental yield is the amount of product that was obtained in the laboratory.

$$\text{percent yield} = \frac{\text{experimental yield}}{\text{theoretical yield}} \times 100\%$$

6. What would be the percent yield if a student made 6.5 grams of product in the lab and theoretically should have made 7.2 grams of product? SHOW YOUR WORK

Name_____

Self-Test 4.6 Calculations with Stoichiometry

SHOW YOUR WORK

1. The mixture of gasoline, C_8H_{18} and oxygen in the fuel injection system of a car is very important to the performance of the engine.

    a) How many moles of O_2 completely react with 6.14 moles (about 1/4 of a gallon) of gasoline?

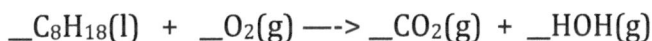

    $$\_C_8H_{18}(l) \ + \ \_O_2(g) \ \text{---> } \_CO_2(g) \ + \ \_HOH(g)$$

    b) How many moles of CO_2 are exhausted into the atmosphere when 6.14 moles of gasoline are burned?

    c) How many grams of CO_2 is this?

2. TUMS™ tablets, $CaCO_3$, are used to neutralize excess stomach acid, HCl. How many grams of calcium carbonate will neutralize 0.100 moles HCl?

    $$\_CaCO_3(s) \ + \ \_HCl(aq) \ \text{--> } \_CaCl_2(aq) \ + \ \_HOH(l) + \_CO_2(g)$$

3. Foam producing fire extinguishers are used for fighting gasoline fires. The foam produced helps smother the fire by preventing oxygen (needed to burn fuel) from entering the system. Determine the mass of CO_2 produced when 250.0 grams of baking soda, $NaHCO_3$ react with excess aluminum sulfate ($Al_2(SO_4)_3$).

$$\_Al_2(SO_4)_3(s) \;+\; \_NaHCO_3(aq) \;—\!\!> \_Al(OH)_3(s) \;+\; \_CO_2(g) \;+\; \_Na_2SO_4(aq)$$

4. Solid carbon undergoes combustion with oxygen gas to produce carbon dioxide. Assuming that the required activation energy is available, write the balanced equation for the reaction between carbon and oxygen, and complete the reaction in the box below showing the correct number of molecules of product (carbon dioxide) and any excess reactant remaining.

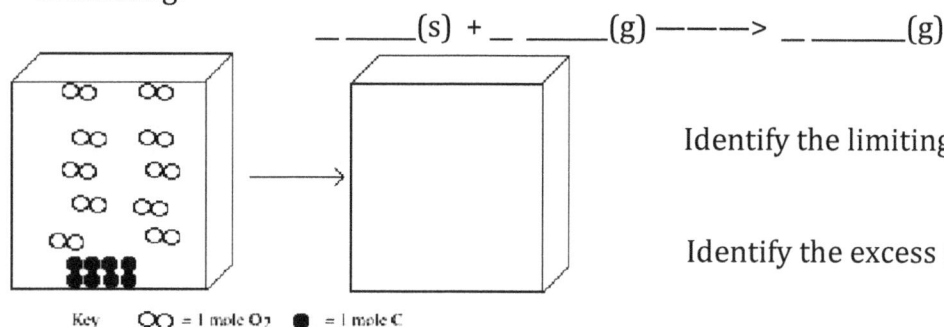

$$\_\_\_\_\_(s) + _____(g) \;—\!\!-\!\!> _____(g)$$

Key ∞ = 1 mole O_2 ● = 1 mole C

Identify the limiting reactant: _____

Identify the excess reactant: _____

5. Solid sulfur and oxygen gas react in the closed container, shown below, to make gaseous sulfur trioxide. Write and then balance the equation for the reaction between sulfur solid and oxygen gas to produce gaseous sulfur trioxide:

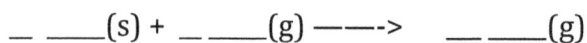

$$\_\_\_\_\_(s) + \_\_\_\_\_(g) \;—\!\!-\!\!> \_\_\_\_\_(g)$$

Complete the reaction between solid sulfur reacting with oxygen gas in the box below.

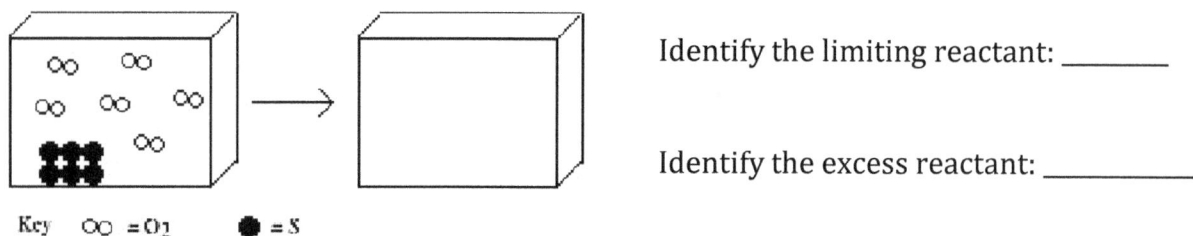

Key ∞ = O_2 ● = S

Identify the limiting reactant: _____

Identify the excess reactant: _____

Name_____

Review Unit 4

1. Calculate the following: SHOW WORK

Substance	Moles	Mass	Total molecules	Total atoms
Fe	1.50			
H_2O	1.50			

2. If a 3.557 g TUMS tablet is 100% $CaCO_3$, how many molecules of $CaCO_3$ make up the tablet? SHOW ALL WORK, LABELS AND UNITS FOR FULL CREDIT.

1. If Ms. Mixon would like to put 1.25×10^{22} molecules of sugar ($C_{12}H_{22}O_{11}$) into her coffee. She has asked you to get it for her. How will you measure this so that she gets the exact amount? SHOW ALL WORK, LABELS AND UNITS FOR FULL CREDIT.

2. Mark T if the statement is true and F if it is false.
   a. _____A mole of Li weighs more than a mole of Cu.

   b. _____ A mole of Zn contains the same number of atoms as a mole of Cu.

   c. _____ A mole of O_2 gas contains more molecules than a mole of CO_2 gas.

   d. _____ A mole of Cl_2 gas contains the same number of atoms as a mole of gold.

3. Iodine is an element that is essential to proper thyroid function. It was common to be deficient in iodine before the addition of sodium iodide to salt (iodized salt). The FDA recommends about 50 mg of iodine per day. Based on percent composition, what is the mass of iodine in a 1.00 g sample of sodium iodide? **SHOW YOUR WORK**

4. Excess oxygen gas reacts with 56.0 grams of aluminum solid to produce aluminum oxide.
    a) How many grams of aluminum oxide will be produced? SHOW YOUR WORK.

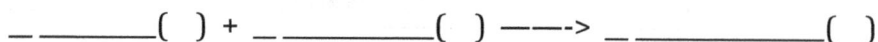

$$\_\ \underline{\hspace{2cm}}(\ \ ) + \_\ \underline{\hspace{2cm}}(\ \ ) ----> \_\ \underline{\hspace{3cm}}(\ \ )$$

    b) How many grams of oxygen are required to completely oxidize the 56.0 g of aluminum?
        SHOW YOUR WORK.

5. Calcium metal reacts with oxygen gas to produce solid calcium oxide:

    a) Balance the equation for the reaction between solid calcium and oxygen gas to produce
        calcium oxide:

$$\_\_Ca\ (s) + \_\_O_2\ (g) -----> \_\_\ \underline{\hspace{2cm}}(\ \ )$$

    b) Assuming that the required activation energy is available, complete the reaction in the
        box below showing the products:

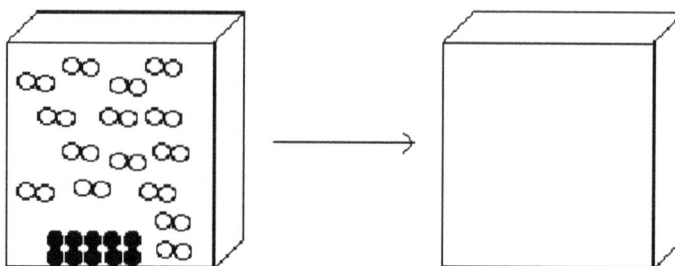

    c) Identify the limiting reagent: _____

    d) Identify the excess reagent: _____

Name_____

6. The reaction of gas A with gas B is represented in the following diagram. Which of the following reactions represents the equation for the reaction that has occurred between A and B.

A= ■
B= ○

        a. $A + 2B \rightarrow AB_2$
        b. $A + B_2 \rightarrow AB_2$
        c. $8A + 3B \rightarrow 3AB + 2B$
        d. $3A + 6B \rightarrow 3 AB_2 + 2B$
        e. none of these

7. In question 8, the substance that is the limiting reagent is:
        a. A
        b. B
        c. neither

8. Calculate the mass of H_2O released into the atmosphere when 15.0 gallons (368 moles) of diesel fuel, $C_{16}H_{34}$, is completely combusted in the engine of a diesel-run vehicle. SHOW YOUR WORK.

        ___ $C_{16}H_{34}(l)$ + ___ $O_2(g)$ —-> ___ $CO_2(g)$ + ___ $H_2O(g)$ + heat

Unit 5
Solution Chemistry

Did you know that?

Molecules have a three-dimensional structure that influences their interactions with other molecules.

Water is the only substance on the surface of Earth that occurs naturally in three different physical states: solid, liquid, and gas.

Rain falls in drops and water beads up on a surface because of hydrogen bonds that occur between water molecules.

Water is one of the few substances that is less dense in the solid state than in the liquid state.

All snowflakes are hexagons.

Without hydrogen bonding, life would be impossible.

Objectives for Unit 5

- I can determine the molecular shape and polarity of a substance when given the chemical name.

- I can use the terms solute and solvent when describing a solution.

- I can draw a molecular model of water molecules at room temperature that shows the polarity of the water molecules and hydrogen bonding between molecules.

- I can draw and explain intermolecular attractions between water and an ionic compound (ion-dipole) and water and a polar molecular solute (dipole-dipole or hydrogen bonding).

- I can determine the solubility of a substance in water and if it is an electrolyte when given the chemical name or formula for the substance.

- I can write a balanced solubility equation for polar solutes in water.

- I can calculate the number of grams of solute needed to make an aqueous molar solution and explain how to make the solution using the most precise method.

- I can perform dilution calculations and explain how to prepare a dilution from a stock solution.

Self-Test 5.1 Molecular Geometry

1. Fill in the following table.

Chemical Name (Chemical Formula)	Lewis Structure	Molecular Shape	Molecule Polarity
hydrogen sulfide ( )			
ammonia (NH_3)			
methane ( )			
carbon tetrachloride ( )			
phosphorus trichloride ( )			
sulfur trioxide ( )			
ethane ( )			
carbon dioxide ( )			
ethyne ( )			
methanol (CH_3OH)			
nitrogen triiodide ( )			

2. For each of the polar covalent molecules, draw the structural formula and determine which atoms have a partial positive charge (**δ+**) and which atoms have a partial negative charge (**δ-**). Make sure to draw unshared pairs of electrons in the structure.

a) HCl

b) H_2O

c) NH_3

d) HBr

Name_____

Self-Test 5.2 Intermolecular Forces in Aqueous Solutions

1. A solution is a homogeneous mixture consisting of a _____ that is dissolved by a _____. An _____ solution is one in which water is the solvent. _____ is considered the _____ solvent because it dissolves more substances than any other.

2. In the box below, draw a molecular view of five water molecules in the liquid phase. Use the following key:

Feature	Symbol
Oxygen atom	●
Hydrogen atom	O
covalent bond	—
hydrogen bond	------

Water molecules at 23°C

3. In the box below, draw a molecular view of what happens when water is mixed with an ionic compound. Show the water molecules and the ionic compound. Correctly orient the water molecules to show the attraction with the negatively and positively charged ions.

$$HOH(l) + NaCl(s)$$

4. What is the name of the intermolecular bond that holds the water molecules to the ion?

5. In the box below, draw a molecular view of what happens when water is mixed with a polar solute, such as methanol. Show the water molecules and the polar solute. Correctly orient the water molecules to show the attraction with methanol. Use the correct molecular structure for methanol in the drawing.

$$HOH(l) + CH_3OH(l)$$

6. What type of intermolecular bond holds the water molecules to the methanol?

7. Determine if the following molecules are soluble in water. If the molecule is soluble, determine if it is an electrolyte. You will need to know the shape of covalent molecules in order to determine their polarity.

Substance	Type (ionic, polar molecular, or nonpolar molecular)	Soluble in water or Insoluble in water	Electrolyte or Nonelectrolyte
C_5H_{12}			
CH_3COOH			
$CuCl_2$			
NH_3			
CCl_4			
$AgNO_3$			

8. Write and balance the following solubility equations:

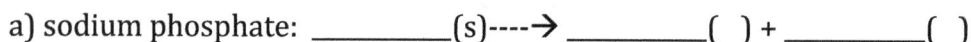

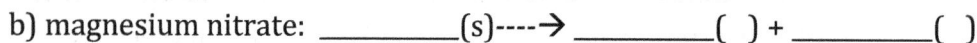

Example: aluminum chloride: $AlCl_3(s) -----\rightarrow Al^{3+}(aq) + 3Cl^-(aq)$

a) sodium phosphate: _____(s)----$\rightarrow$ _____( ) + _____( )

b) magnesium nitrate: _____(s)----$\rightarrow$ _____( ) + _____( )

c) 2-propanol: _____(l)----$\rightarrow$ _____( )

Name_____

Self-Test 5.3 Molarity

Molarity = <u>moles of solute</u>   Unit: M or mol/L
            liter of solution

<u>To prepare a solution from a solid:</u>
1. Determine the molarity of the solution and the required volume.
2. Use the equation, M x V = moles (volume must be in Liters) to calculate the number of moles needed for the solution.
3. To find the number of grams of solute needed, convert the moles of solute to grams using the molecular mass of the solute.

<u>To prepare a dilution from a prepared solution:</u>
1. Determine the molarity of the original solution (M_1) and the molarity (M_2) and volume (V_2) of the resulting solution.
2. Use: $M_1V_1 = M_2V_2$ to solve for the volume of concentrated solution needed (V_1).

SHOW YOUR WORK AND LABELS FOR FULL CREDIT
1. Lye (NaOH) is an effective drain cleaner. Describe how to prepare 25.0 mL of 1.35 M NaOH, starting with solid NaOH. Make sure to use the correct names of the equipment used to make a solution and precise directions.

2. A hydrate of sodium thiosulfate, known as hypo ($Na_2S_2O_3$), is used as a fixer in photography because it dissolves silver compounds. Show how to prepare 250.00 mL of 0.4500 M solution of hypo from solid hypo. Make sure to use the correct names of the equipment used to make a solution and precise directions.

3. What is the molar concentration of a solution in which 0.855 moles of washing soda (Na_2CO_3) is dissolved in water to make 500.0 mL of a solution for softening water.

4. In car batteries, 6.25 M H_2SO_4 (sulfuric acid) is used. If 18.0 M sulfuric acid is purchased, how could the concentrated acid be diluted to get the right molarity if 3.50 L of the 6.25M battery acid is needed? Make sure to use the correct names of the equipment used to make the solution and precise directions.

5. One of the uses of methanol (CH_3OH) in dilute form is as windshield washer antifreeze. In pure form, methanol has a molar concentration of 24.7 M. How would 2.50 L of a 6.45M solution of methanol be prepared? Make sure to use the correct names of the equipment used to make a solution and precise directions.

6. A sample of household ammonia (NH_3) contains 46.6 g of NH_3 gas dissolved in water to form 250.0 mL of solution. What is the molarity of the household ammonia?

Name_____

Review Unit 5

1. Fill in the table:

Substance	Lewis Structure	Molecular Shape	Polarity
NH_3			
H_2S			
CH_3OH			
I_2			
C_2H_4			

2. Show the solubility equation.

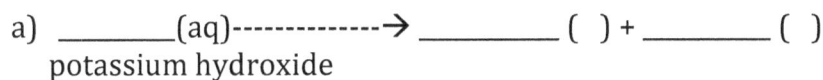

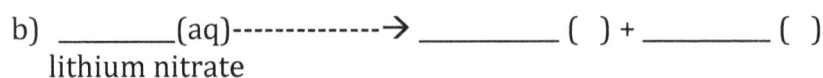

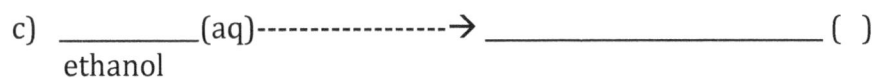

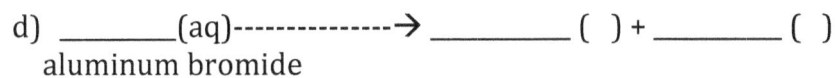

    a) _____(aq)-------------→ _____ ( ) + _____ ( )
        potassium hydroxide

    b) _____(aq)-------------→ _____ ( ) + _____ ( )
        lithium nitrate

    c) _____(aq)-----------------→ _____ ( )
        ethanol

    d) _____(aq)-------------→ _____ ( ) + _____ ( )
        aluminum bromide

3. Calculate the number of grams of solid NaOH that you will weigh out to make 25.00 mL of 2.450 M solution.

4. What is the molarity of a solution that contains 45.31 g of Na_2CO_3 dissolved in 250.00 mL of solution?

5. If you purchase 12.0 M battery acid (H_2SO_4) in order to make 750.00 mL of a 3.50 M aqueous solution, what volume of the 12.0 M battery acid will be used?

6. Mark A if the following dissolves in water and B if it does not:
   a) _____ NaOH

   b) _____ $Cu(NO_3)_2$

   c) _____ CH_3OH

   d) _____ $C_6H_{12}O_6$ (glucose)

   e) _____ CH_4

   f) _____ C_6H_{14}

7. Mark A if the substance is an electrolyte and B if it is a nonelectrolyte.
   a) _____ Na_2SO_4

   b) _____ KCl

   c) _____ CH_3OH

   d) _____ CCl_4

Unit 6
Chemical Reactions

Did you know that?

Six of the top-ten chemicals produced in the U.S. are either acids or bases.

Lemon neutralizes bad smelling and bad tasting amines in fish.

Swimmers can experience acidosis where blood pH is below normal.

Basic shampoos break disulfide bonds in hair, causing split ends.

Alcohol can damage the membrane that protects your stomach from the digestive acid, hydrochloric acid.

Neutralization reactions are responsible for the destruction of famous statues.

Fluorides in toothpastes convert tooth enamel to an acid-resistant fluoride compound.

Silver tarnishes due to an oxidation-reduction reaction

Gold is found in nature as a pure solid because its potential to be oxidized into an ore is so low.

The San Shi Liu, written in the 6th century AD, contains detailed instructions for making batteries.

Objectives for Unit 6

- I can write and balance formula equations, complete ionic equations, and net ionic equations when given a written precipitation reaction.

- I can use the solubility rules to predict the solubility of an ionic compound and the formation of a precipitate in a chemical reaction.

- I can draw a molecular model to show spectator ions and the formation of a precipitate in an aqueous solution.

- I can calculate the amount of precipitate formed when given the molarity and volume of two ionic solutions.

- I can calculate the quantity of a reactant needed to fully precipitate an ion from a solution when given the molarity and volume of the initial solution.

- I can write dissociation equations for strong acids and bases and define what a strong acid and base are.

- I can calculate pH, pOH, [H+], and [OH-], for a solution when given one of the values.

- I can write and balance neutralization reactions when given the chemical name for the acid and base reactants or the products.

- I can calculate the amount of acid needed to neutralize a certain molarity and volume of a base.

- I can use data from a titration to determine the concentration of an unknown acid or base.

- I can write and balance oxidation/reduction reactions and determine the substance that is oxidized and reduced.

- I can write balanced half-reactions for oxidation and reduction and determine the oxidizing agent, reducing agent, and spectator ions.

Name_____

Self-Test 6.1 Precipitation Reactions

A precipitate is a solid substance that forms when <u>two aqueous solutions</u> are mixed together. Not all aqueous solutions that are mixed will form a precipitate. The precipitate occurs when ions from each solution are so attracted to each other when they are mixed that the water molecules cannot keep them apart.

Example:

$$Pb(NO_3)_{2(aq)} + 2KI_{(aq)} ----\rightarrow PbI_{2(s)} + 2KNO_{3(aq)}$$

aqueous lead(II) nitrate + aqueous potassium iodide ----→ solid lead(II) iodide + aqueous potassium nitrate

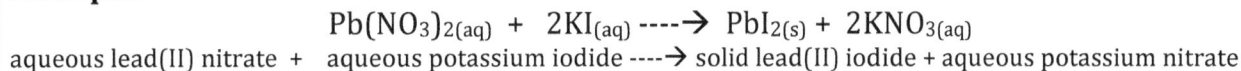

complete ionic equation:

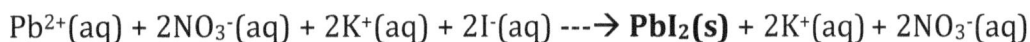

$$Pb^{2+}(aq) + 2NO_3^-(aq) + 2K^+(aq) + 2I^-(aq) ---\rightarrow \textbf{PbI}_2\textbf{(s)} + 2K^+(aq) + 2NO_3^-(aq)$$

1. Draw a molecular model of an aqueous solution of lead (II) nitrate in beaker 1, an aqueous solution of potassium iodide in beaker 2, and the resulting product when the two solutions are mixed in beaker 3. Remember to include the ions and the water molecules in all the beakers. Use the following symbols:

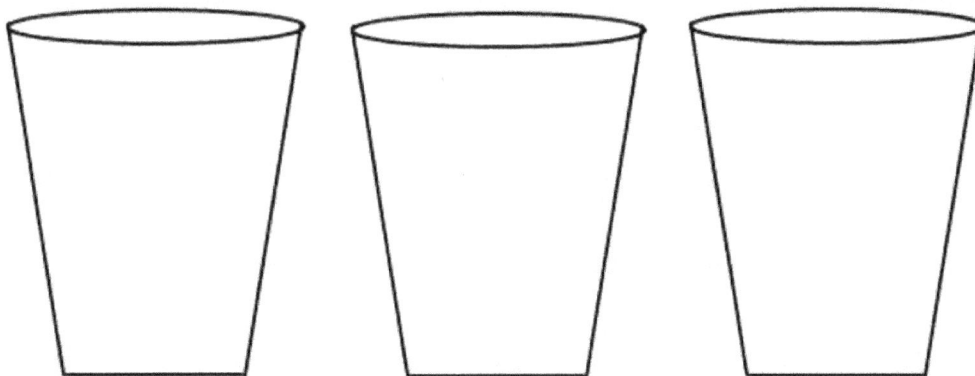

Pb^{2+} ▲ NO_3^- ● K^+ ◆ I^- ○

       lead(II) nitrate(aq)   potassium iodide(aq)   solution 1 + solution 2

2. Mark the ions in the final mixture that are the **spectator ions**.
   a) NO_3^-
   b) K^+
   c) Pb^{2+}
   d) I^-

3. Fill in the missing information in the final **net ionic equation** below:

net ionic equation: _____(aq) + _____(aq) ---→ PbI_2(s)

4. In the table below, write correct formulas for ionic compounds using the combination of ions noted. Use the solubility rules to predict which of the resulting compounds is soluble in water (aq) or insoluble in water (s).

Ion	nitrate	sulfate	hydroxide
lead(II)			
calcium			
silver			

Write out a balanced chemical equation for each of the problems below. Make sure your formulas are correct and show the individual ions and charges as seen in the example problem above.

5. The reaction between aqueous lithium hydroxide and aqueous lead (IV) nitrate.
a) formula equation

b) complete ionic equation

c) net ionic equation

6. The reaction between aqueous sodium sulfide and aqueous nickel (III) nitrate.
a) formula equation

b) complete ionic equation

c) net ionic equation

7. The reaction between aqueous aluminum sulfate and aqueous barium nitrate.

d) formula equation

e) complete ionic equation

f) net ionic equation

Name_____

Self-Test 6.2 Stoichiometry of Precipitation Reactions

1. 50.0 mL of a 0.100M solution of aqueous silver nitrate is mixed with 25.0 mL of a 0.500M solution of aqueous copper(II)chloride. What mass of solid silver chloride will be produced?

a) Write the chemical equation for the reaction and balance.

b) Determine the number of moles of each reactant.

c) Determine the limiting reactant.

d) Use the limiting reactant to determine the number of moles of solid silver chloride will be produced.

e) Convert the number of moles of silver chloride to grams.

2. 0.500L of a 0.50M solution of sodium hydroxide is mixed with 0.250L of 1.00M aqueous iron(II) chloride to form a green precipitate. Determine the identity and mass of the resulting precipitate.

3. In order to properly dispose of a barium solution (a heavy metal), the barium must be precipitated out of solution and disposed of as hazardous waste. Determine the amount of solid sodium phosphate needed to precipitate all of the barium ions from 1.00L of 0.25M aqueous barium nitrate solution.

   a) Write the chemical equation for the reaction and balance.

   b) Determine the number of moles of barium nitrate present.

   c) Use the mole to mole ratio of sodium phosphate to barium nitrate to determine the number of moles of sodium phosphate needed for the reaction.

   d) Convert the moles of sodium phosphate to grams.

Name_____

Self-Test 6.3 Properties of Acids and Bases

1. Fill in the table.

Formula	Name	Use
	phosphoric acid	fertilizer production
H_2CO_3		carbonated beverages
H_2SO_4		battery acid
H_2SO_3		acid rain
	hydrofluoric acid	etches glass
$Ca(OH)_2$		plaster

2. Strong acids and bases dissociate completely when mixed with water. Write the dissociation equations for each of the following acids or bases.

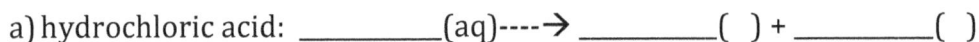

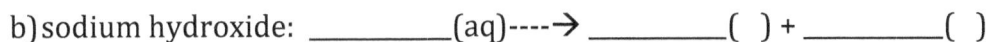

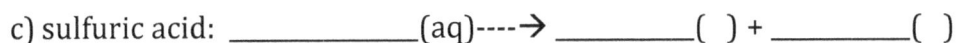

   a) hydrochloric acid: _____(aq)---->_____( ) + _____( )

   b) sodium hydroxide: _____(aq)---->_____( ) + _____( )

   c) sulfuric acid: _____(aq)---->_____( ) + _____( )

   d) calcium hydroxide: _____(aq)---->_____( ) + _____( )

pH

Formulas: **pH = -log [H⁺]**   **pOH = -log [OH⁻]**   **pOH + pH = 14**   **[H⁺] [OH⁻] = 1.00 x 10⁻¹⁴**

3. Fill out the following table.

Substance	[H+]	[OH-]	pH	pOH	acid or base?
pumpkin pulp	1.0×10^{-5} M				
urine	5.0×10^{-9} M				
saliva			7.0		
beer		3.1×10^{-10} M			
soda pop	1.0×10^{-3} M				
stomach contents			2.0		
vinegar			2.9		
milk				7.5	
apples		1.26×10^{-11} M			
blood	3.98×10^{-8} M				
intestinal contents			6.5		

4. Complete the dissociation equation for these strong acids and bases. Note the molarity of the [H⁺] or [OH⁻] produced, and then calculate the pH. SHOW YOUR WORK.

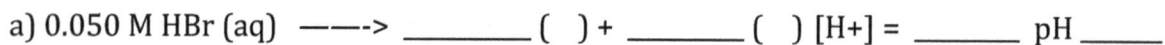

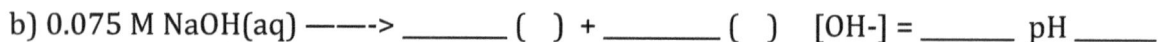

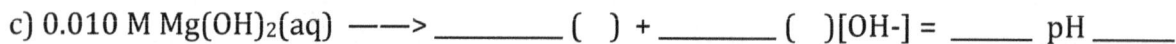

a) 0.050 M HBr (aq) –––-> _____ ( ) + _____ ( ) [H+] = _____ pH _____

b) 0.075 M NaOH(aq) –––-> _____ ( ) + _____ ( ) [OH-] = _____ pH _____

c) 0.010 M Mg(OH)₂(aq) –––> _____ ( ) + _____ ( )[OH-] = _____ pH _____

5. Calculate the following: SHOW YOUR WORK and LABELS.

a) pH of a 3.27 x 10⁻⁵ M solution of NaOH.

b) [OH⁻] of a solution with a pH = 2.55.

Self-Test 6.3 Neutralization Reactions

1. Write and balance the neutralization reactions for the following:

   a) sulfuric acid and aqueous potassium hydroxide

   b) nitric acid and aqueous calcium hydroxide

   c) hydrobromic acid and aqueous lithium hydroxide

   d) hydrochloric acid and aqueous sodium hydroxide

Acid Base Stoichiometry and Titration

2. What is the molarity of a solution of NaOH if 20.0 mL of the NaOH are required to neutralize 12.0 mL of 2.00 M H_2SO_4?

   a) Write a balanced equation that shows the stoichiometry between the acid and the base.

   ___ _____ + ___ _____ ——-> ___ _____ + ___ _____

   b) Calculate the number of moles of H_2SO_4. Use **M x V = moles**. SHOW YOUR WORK.

   c) Use the mole to mole ratio to determine how many moles of NaOH were neutralized by the acid. SHOW YOUR WORK.

   d) Calculate the molarity of the base. **Use moles/L = M**. SHOW YOUR WORK.

3. A 0.100 M sodium hydroxide solution was used to titrate a solution of hydrochloric acid of unknown concentration. At the endpoint, 21.0 mL of NaOH solution had neutralized 100.0 mL of the hydrochloric acid solution. What is the molarity of the HCl solution? SHOW YOUR WORK.

4. A 0.150M KOH solution fills a buret to the 0.00 mark. The solution was used to titrate 25.00 mL of an HNO_3 solution of unknown concentration. At the endpoint, the buret reading was 34.60 mL. What is the molarity of the HNO_3 solution? SHOW YOUR WORK.

Name_____

Self-Test 6.4 Redox Reactions

1. When a zinc strip is placed in a solution of copper (II) nitrate, copper solid and aqueous zinc nitrate are formed. Write and balance the full redox reaction and the half reactions. Then, identify the oxidizing agent, the reducing agent, and the spectator ion.

a) Full redox reaction:

b) Oxidation half reaction: The substance that is oxidized: _____

c) Reduction half reaction: The substance that is reduced: _____

The oxidizing agent: _____   The reducing agent: _____   The spectator ion: _____

2. When a square of aluminum foil is placed in copper (II) chloride solution, solid copper and aqueous aluminum chloride are formed. Write and balance the full redox reaction and the half reactions. Then, identify the oxidizing agent, the reducing agent, and the spectator ion.

a) Full redox reaction:

b) Oxidation half reaction: The substance that is oxidized: _____

c) Reduction half reaction: The substance that is reduced: _____

The oxidizing agent: _____   The reducing agent: _____   The spectator ion: _____

3. Aqueous hydrochloric acid and solid magnesium combine to form aqueous magnesium chloride and hydrogen gas. Write and balance the full redox reaction and the half reactions. Then, identify the oxidizing agent, the reducing agent, and the spectator ion.

a) Full redox reaction:

b) Oxidation half reaction: The substance that is oxidized: _____

c) Reduction half reaction: The substance that is reduced: _____

The oxidizing agent: _____   The reducing agent: _____   The spectator ion: _____

4. The number of electrons lost during oxidation must always be equal to:
    a) the charge on the ion.
    b) the number of electrons gained in reduction.
    c) the number of electrons gained by the oxidizing agent.
    d) the total change in oxidation numbers.

5. For each of the following metals, predict if a reaction will occur when the metal is placed in 1.0 M solution of tin (II) chloride. Use the Activity Series to make predictions. Write R for reaction and NR for no reaction.
    a) ____ solid zinc

    b) ____ solid magnesium

    c) ____ solid copper

Name_____

Review Unit 6

1. Complete and balance the reaction and identify the following

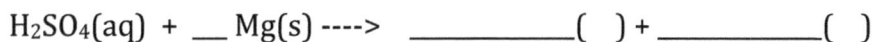

    $H_2SO_4(aq)$ + __ $Mg(s)$ ----> _____( ) + _____( )

    a) oxidation half-reaction:

    b) reduction half-reaction:

    c) oxidizing agent: _____

    d) reducing agent: _____

    e) spectator ion _____

2. If 5.00 grams of copper solid react with excess aqueous silver nitrate to produce aqueous copper(II) nitrate and silver solid, how many grams of silver solid will be produced? SHOW YOUR WORK and LABELS.

    _ _____ ( ) + _ _____( ) ——> _ _____ ( ) + _ _____( )

    b) For the reaction in problem 2 identify:

    ____a) the substance that is being oxidized   A. Ag^+

    ____b) the oxidizing agent   B. Cu^{2+}

    ____c) the substance that is being reduced   C. Ag^0

    ____d) the reducing agent   D. NO_3^-

    ____e) the spectator ion   E. Cu^0

3. Excess oxygen gas reacts with 56.0 grams of aluminum solid to produce aluminum oxide. How many grams of aluminum oxide will be produced? SHOW YOUR WORK.

$$\underline{\ \ }\underline{\ \ \ \ \ \ \ \ \ }(\ \ ) + \underline{\ \ }\underline{\ \ \ \ \ \ \ \ \ }(\ \ ) ----> \underline{\ \ }\underline{\ \ \ \ \ \ \ \ \ \ \ }(\ \ )$$

b) How many grams of oxygen are required to completely oxidize 56.0 g of aluminum? SHOW YOUR WORK.

c) For the reaction in problem 3 identify:

_____1) the substance that is being oxidized   A. Al^{3+}

_____2) the oxidizing agent   B. O^{2-}

_____3) the substance that is being reduced   C. Al^{0}

_____4) the reducing agent   D. O_2^{0}

Write the following equations for the precipitation reactions.

4. The reaction between aqueous barium nitrate and aqueous sodium sulfate.
   a) formula equation

   b) complete ionic equation

   c) net ionic equation

Name_____

5. The reaction between aqueous silver nitrate and aqueous sodium phosphate.
   a) <u>formula equation</u>

   b) <u>complete ionic equation</u>

   c) <u>net ionic equation</u>

6. Fill in the table for acids and bases.

Formula	Name
	calcium hydroxide
	phosphoric acid
H_2CO_3	
	aluminum hydroxide
	hydrosulfuric acid
H_2SO_4	

7. Calculate the following: SHOW YOUR WORK

a) Calculate [H+] when the pH = 4.5   b) Calculate [OH-] when the pH = 9.2

c) Calculate [OH-] when [H+] is $2.2 \times 10^{-2}M$ d) Calculate pH when [H+] = $3.0 \times 10^{-4}M$

8. The process of adding standard solution from a buret in controlled amounts is called

   _____. The theoretical point at which the number of moles of acid equals

   the number of moles of base is called the _____ point and is visualized at

   the _____ when the _____ changes color.

9. Complete the following neutralization reactions

   a) $\_Ba(OH)_2(aq)$ + $\_HCl(aq)$ ——-> _____( ) + _____( )

   b) $\_HCl(aq)$ + $\_KOH(aq)$ ——-> _____( ) + _____( )

   b) $\_H_2SO_4(aq)$ + $\_LiOH(aq)$ ——-> _____( ) + _____( )

10. In a titration of 45.0 mL of a sulfuric acid solution, the end point is reached when 35.0 mL of 0.100M sodium hydroxide is added. Calculate the concentration of the sulfuric acid solution. SHOW YOUR WORK.

Did you know that?

Atoms within a substance are always moving unless the temperature reaches absolute zero.

It is harder to breathe at the top of a mountain because there is less atmosphere to push the air into the lungs.

The cliché that a person who has a great responsibility "bears the weight of the world" would more accurately be expressed as "bears the weight of the atmosphere".

Gases that are trapped in the middle ear can cause painful pressure on the eardrum during takeoff and landing in a plane, which makes passengers want to 'pop' their ears.

Objectives for Unit 7

- I can describe relationships between kinetic energy, temperature, and motion of particles.

- I can draw a molecular model for a substance in the solid, liquid, and gas state.

- I can describe pressure, atmospheric pressure, and identify different units of measurement used for pressure.

- I can describe the conditions of STP and use the standard molar volume of a gas to determine how much volume a certain amount of gas will take up at STP.

- I can use Boyle's, Charles', Gay-Lussac's, and the Ideal Gas Law to calculate and solve problems related to ideal gases.

- I can solve gas stoichiometry problems in chemical reactions at STP using the molar volume of a gas.

Name_____

Self-Test 7.1 Properties of Gases

1. Bromine (Br_2) exists as a liquid at room temperature (23°C or 296K). The freezing point of bromine is 265.95K and the boiling point is 331.95K. In each of the boxes below, draw 9 molecules of bromine (Br_2) at the specified temperatures:

Br_2= ⬤⬤

Box A	Box B	Box C

200K	300K	400K

2. Answer the following questions by choosing A (Box A), B (Box B), or C (Box C).

   a) _____ Contains the molecules with the greatest average kinetic energy.

   b) _____ Contains molecules that can be compressed.

   c) _____ Contains the state of bromine that has the highest density.

   d) _____ Contains the state of bromine with a definite shape.

   e) _____ Container with the greatest pressure.

   f) _____ Container that has the greatest distance between particles.

3. Mark each of the following relationships as direct (D), indirect (I), or no relationship (N).

   a) _____ The temperature of a gas in a closed container and the pressure.

   b) _____ The number of particles of gas in a closed container and the pressure.

   c) _____ The kinetic energy of gas particles in a closed container and the temperature.

4. T/F If it can, a gas will move from an area of low pressure to an area of high pressure.

    a) Based on your answer to the question #4 above, give an example that demonstrates this concept.

5. Fill in the following table regarding absolute zero.

Absolute Zero	
Temperature in °C	
Temperature in K	
Kinetic energy of the particles in a substance	
Motion of the particles in a substance	
Pressure at absolute zero	

Gas Pressure and the Atmosphere

6. Pressure is caused by gas particles _____

    _____.

7. There are several units that scientists use to measure gas pressure. Atmospheric pressure at sea level is 1 atmosphere (1 atm). List the equivalent measurement in the following units:

    a) _____mm Hg   (common unit in mercury barometer)

    b) _____kPa   (commonly used for small differences in gas pressure)

    c) _____psi   (common unit for tire pressure in English system)

8. Explain why the atmospheric pressure is lower at the top Mt. McKinley than on Waikiki Beach in Honolulu, Hawaii. Draw a diagram to illustrate your answer.

9. List the conditions of STP: Temperature _____°C   _____K

                                Pressure   _____atm   _____mm Hg

Name_____

Self-Test 7.2 Gas Relationships for Ideal Gases

Refer to the following Gas Laws:

Boyle's Law: $P_1V_1=P_2V_2$ **Charles's Law: $\dfrac{V_1}{T_1}=\dfrac{V_2}{T_2}$**   **Gay-Lussac's Law: $\dfrac{P_1}{T_1}=\dfrac{P_2}{T_2}$**

*For gas laws that include temperature, the temperature must be measured in K

WRITE OUT ALL KNOWNS, INCLUDE THE FORMULA, SHOW YOUR WORK, INCLUDE UNITS

1. The pressure in a sealed can of pop is 120.0 kPa at 4°C. What will the pressure be when the can is left at room temperature (23°C)?

2. 1.00L of a gas at STP is compressed to 473 mL. What is the new pressure of the gas?

3. The temperature outside in January averages -3.5°C. A happy toddler carries a helium balloon with an initial volume of 6.5 L from inside a store (temperature of 22°C) to the parking lot outside. Circle the face you would expect to see on the toddler.

   a)

   b) Calculate the resulting volume of the balloon.

   c) What will you tell the toddler to assure her that the balloon can be fixed?

4. The pressure in a bike tire is 520 kPa at 21°C. After biking on hot pavement, the tire heats up to 37°C. Assuming that the volume of the gas remains constant, what is the resulting tire pressure?

5. The air in an oven is 23.5°C and 101.3 kPa. If the oven remains closed and is then turned up to 260°C, what is the resulting pressure in the oven? What will happen when the door is opened to put food in?

Name_____

Ideal Gas Law: **PV=nRT**   **R= 0.0821L•atm/mol•K**

Solve the following problems using the Ideal Gas Law. SHOW WORK, INCLUDE UNITS ON ALL NUMBERS, AND USE CORRECT SIG FIGS.

6. What is the number of moles of gas in a standard automobile tire with a volume of 10.0L at a pressure of 2.72 atm and a temperature of 5°C?

7. At sea level, at a temperature of 23°C, the average lung capacity of a person is 6 L.
   a) Calculate the number of moles of air the average person can inhale in one breath.

   b) If air contains only about 21% oxygen, how many moles of oxygen is this?

   c) Mountain climbing requires a technique called pressure breathing in order to combat the decreasing pressure as altitude increases. The atmospheric pressure on the top of Mount Everest is about 0.333 atm and the temperature in the summer can be a chilly -14°C. The average lung capacity is still 6 L, how many moles of gas can a person inhale in one breath under these conditions?

   d) The percentage of oxygen in the atmosphere at the top of Everest is still 21%, how many moles of oxygen will each breath contain on the mountain?

8. Pressure breathing requires that the person uses the diaphragm muscles to force the carbon dioxide out of the lungs by fully relaxing the muscles and pull air into the lungs by fully contracting the diaphragm. What property of the ideal gas law does this change in order to improve effective oxygen and carbon dioxide exchange?

Standard Molar Volume of an Ideal Gas:

At STP, one mole (6.022 x 10²³ particles) of gas has a volume of 22.4 L. This applies to all gases.

9. Two gas cylinders have the same volumes and are both at STP, but they hold different gases. The first cylinder holds 1.50 moles of propane (C_3H_8), while the second holds chlorine gas (Cl_2).

Container 1   Container 2

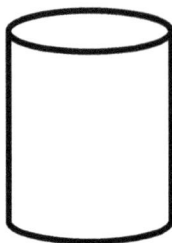   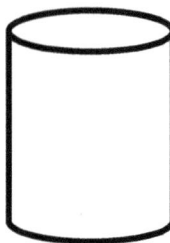

1.5 moles C_3H_8 ? moles Cl_2

Fill in the table below and **include your calculations**.

	Container 1 (C_3H_8)	Cl_2
Moles	1.50	
Molecules		
Atoms		
Mass (g)		
Volume (L)		

b) When comparing gases at STP, which measurements (moles, molecules, atoms, mass, volume) will <u>always</u> be the same for both gases?

c) Which measurements may be different?

Name_____

Stoichiometry Problems Using Molar Volume of an Ideal Gas:

10. In order to power a small bottle rocket at STP, 1.50L of hydrogen gas must be generated from the following reaction:

$$_____ Mg_{(s)} + _____ HCl_{(aq)} \text{ --------} \rightarrow \_\_\_\_ H_{2(g)} + _____ MgCl_{2(aq)}$$

     a) Balance the equation.

     b) Calculate the number of moles of hydrogen gas needed. Use molar volume to solve.

     c) Determine how many grams of magnesium are required for the reaction if excess hydrochloric acid is used.

11. What volume of carbon dioxide gas will be generated when 5.00g of sodium bicarbonate react with excess acetic acid? Assume the conditions are at STP.

$$_____ NaHCO_{3(s)} + _____ CH_3COOH_{(aq)} \text{ --------} \rightarrow _____ NaCH_3COO_{(aq)} + _____ CO_{2(g)} + \_\_\_\_ H_2O_{(l)}$$

     a) Balance the equation.

     b) Calculate the number of moles of sodium bicarbonate available.

     c) Determine the number of moles of carbon dioxide produced.

     d) Use the molar volume to determine the volume of carbon dioxide produced.

Name_____

Review Unit 4

1. Two cylinders are equal in volume and are both at STP but they hold different gases. The first cylinder holds 3.35 moles of carbon dioxide, CO_2 (g), and the second holds helium gas (He).

   Fill in the table below and **include your calculations**.

	Container 1 (CO_2)	Container 2 (He)
Moles	3.35	
Molecules		
Atoms		
Mass (g)		
Volume (L)		

2. Explain why the warning label on an aerosol can states, "do not incinerate". Use Gay-Lussac's Law in your explanation.

3. Why do scuba divers need to equalize the pressure in their ears when descending on a dive? What happens to pressure as the depth increases, and what does this do to the gas in the middle ear?

4. For each of the following questions, write increases or decreases.

     a) Increasing the temperature of a gas in a fixed volume container _____the pressure inside the container.

     b) Decreasing the volume of a container of gas while keeping the temperature constant _____ the pressure inside the container.

     c) Decreasing the temperature of a gas in a container with a movable piston _____ the volume of the container.

     d) Increasing the number of molecules of gas in a fixed container _____ the pressure inside the container.

     e) Decreasing kinetic energy _____ temperature.

5. Yeast catalyzes the breakdown of hydrogen peroxide to form water and oxygen gas. In the presence of excess yeast solution, what volume of oxygen gas will be produced from 100.0 mL of a 1.00M solution of hydrogen peroxide? Assume STP.

$$\underline{\hspace{2cm}} H_2O_{2(aq)} \text{ --------} \rightarrow \underline{\hspace{1cm}} H_2O_{(l)} + \underline{\hspace{1cm}} O_{2(g)}$$

     a) Balance the equation.

     b) Determine the moles of hydrogen peroxide available (M•V=moles).

     c) Calculate the number of moles of oxygen gas produced.

     d) Determine the volume of oxygen produced using the molar volume.

Chapter 8
Intermolecular Bonding and Properties of Substances

Did you know that?

Water boils at different temperatures in Minnesota and Colorado. This affects the cooking of many foods at high elevations.

Geckos employ dispersion bonding to climb.

Freezing point depression makes sodium chloride an effective deicer.

It's possible for nonpolar molecules, such as carbon dioxide, to dissolve in a polar solvent. The soda industry has profited from this.

Some mineral springs contain naturally carbonated water.

Fish and other oxygen dependent organisms rely on oxygen's ability to dissolve in water.

Objectives for Unit 8

- I can identify relationships between temperature, kinetic energy, and amount of intermolecular bonding during phase changes.

- I can use the terms endothermic and exothermic to correctly label phase changes in terms of energy exchange due to intermolecular bond breaking and forming.

- I can label a phase change diagram with melting point, boiling point, solid, liquid, and gas.

- I can explain what happens on a molecular level when a liquid is heated to the boiling point. I can describe the effect of differences in atmospheric pressure on the boiling point.

- I can determine the type of intermolecular bond present in a substance when given the chemical name or chemical formula.

- I can determine the relative strength of the intermolecular bonding within a substance when compared to another substance and relate the type of bond to properties such as melting point and boiling point.

- I can explain how a nonpolar and polar substance dissolves in water in terms of intermolecular bonding and I can list the conditions that increase the solubility of both polar and nonpolar substances.

- I can compare the effect of various polar solutes on the boiling point and freezing point of water when in an aqueous solution.

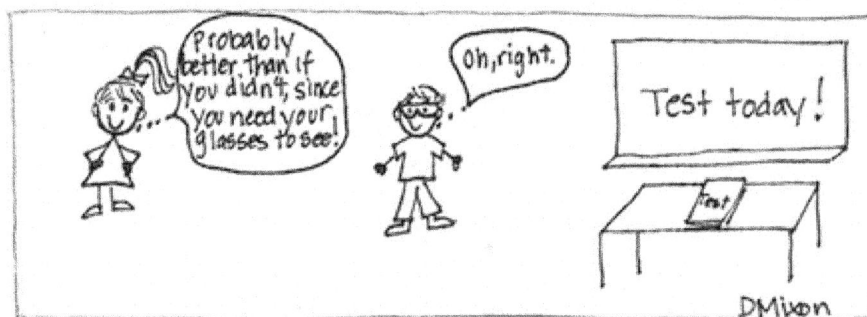

Name_____

Self-Test 8.1 Intermolecular Bonding, Phase Changes, and Energy

1. Draw a diagram of water molecules in the solid, liquid and gas phase. Make sure to show the relative shape and distance of the molecules in each phase. Draw 6 molecules for each.

Feature	Symbol
Oxygen atom	●
Hydrogen atom	○
covalent bond	▬
hydrogen bond	------

| Solid | Liquid | Gas |

2. Recall that kinetic energy is directly related to the temperature of a substance. As the temperature of water increases, kinetic energy _____ (increases or decreases), the motion of the water molecules _____ (increases or decreases), and the number of intermolecular bonds between water molecules _____ (increases/decreases).

3. Energy is required to break intermolecular bonds in a substance and energy is given off when intermolecular bonds are formed. Decide if the phase change is endothermic (A) or exothermic (B) in the following phase changes.

    a) _____ Solid to a liquid.

    b) _____ Gas to a liquid.

    c) _____ Liquid to a solid.

    d) _____ Liquid to a gas.

4. The graph below is a heating curve that shows the relationship between temperature and energy input. Use the graph to answer the following questions:

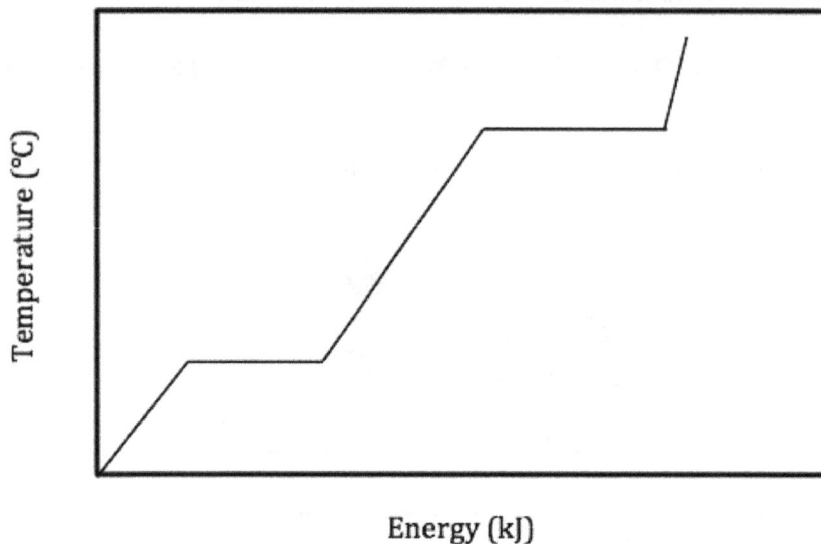

Energy (kJ)

Label the graph with the following information:

   a) Location on the curve where there is the greatest number of intermolecular bonds.

   b) Location on the curve where the substance exists as only a liquid.

   c) Location on the curve where the substance exists as both a liquid and a gas.

   d) Location on the curve where the number of intermolecular bonds is the least.

   e) Location on the y-axis noting the temperature of the melting point.

   f) Location on the y-axis noting the temperature of the boiling point.

Atmospheric Pressure and Boiling Point of a Liquid

5. As the temperature of a liquid increases, the average kinetic energy of the atoms or molecules _____and they begin to move _____. When the particles absorb enough energy, they begin to change from a liquid to a _____and form vapor bubbles. The vapor in the bubbles exerts a pressure and the bubbles _____in size. When the vapor pressure inside the bubbles _____the atmospheric pressure, the liquid boils. The temperature at which a liquid boils is _____related to atmospheric pressure. When the atmospheric pressure is_____, liquids boil at a higher temperature. At low atmospheric pressure, such as in the mountains, liquids boil at a _____temperature. Boiling is different than evaporation because _____ _____ occurs at the heat source, while evaporation occurs at the _____.

Name_____

8.2 Intermolecular Bonding and Physical State

1. Complete the following table. The dashes between the molecules represent intermolecular forces.

Molecules	Molecular mass (of one molecule of the substance)	Physical Phase (found on the periodic table)	Molecule Polarity	Type of Intermolecular force Dipole – dipole Hydrogen bonding Dispersion force
H_2-----H_2				
F_2-----F_2				
O_2-----O_2				
Br_2-----Br_2				
I_2-----I_2				

2. Explain why bromine is a liquid and iodine is a solid when the other substances in problem #1 are gases.

3. Refer to the table below to answer the next questions. Melting and boiling points of saturated hydrocarbons are at standard pressure.

Alkane	Melting Point °C	Boiling Point °C
methane	-183	-164
ethane	-182	-89
propane	-190	-42
butane	-138	-0.5
pentane	-130	36
hexane	-95	69
heptane	-91	98
octane	-57	125
nonane	-51	151
decane	-30	174
eicosane ($C_{20}H_{42}$)	37	343

a) T/F____The alkanes above are nonpolar molecules that are attracted to each other through dispersion forces.

b) T/F____Boiling point is indirectly related to molecular mass of a hydrocarbon.

c) T/F____Boiling point is directly related to the number of carbons in a hydrocarbon.

d) T/F____Butane is a liquid at room temperature (25°C) and standard pressure.

e) T/F____Nonane is a liquid at room temperature (25°C) and standard pressure.

f) T/F____Eicosane is the only hydrocarbon on the list that would be a solid at room temperature (25°C) and standard pressure.

4. Draw and explain how oxygen gas (O_2) dissolves in water. Use the following symbols:

oxygen	water

Name_____

Self-Test 8.3 Intermolecular Bonding and Properties of an Aqueous Solution

Boiling Point Elevation and Freezing Point Depression

1. Arrange the following in order of boiling and freezing points:

    a) 2.0 mL of methanol in 20.0 mL of water   highest b.p._____   highest f.p._____

    b) 4.0 mL of ethanol in 20.0 mL of water   _____   _____

    c) 3.0 mL of 1-propanol in 10.0 mL of water   lowest b.p _____   lowest f.p _____

2. Show the dissociation equation for the aqueous solution and then arrange in order of boiling and freezing points.

    a) 1.0 M $CaCl_2$ ----> _____ ( ) + _____ ( ) highest b.p._____ highest f.p._____

    b) 1.0 M Na_3PO_4 ----> _____ ( ) + _____ ( )   _____   _____

    c) 1.0 M NaCl ----> _____ ( ) + _____ ( )   _____   _____

    d) 1.0 M $C_{12}H_{22}O_{11}$ ----> _____ ( )   lowest b.p. _____ lowest f.p._____

3. Show the dissociation equation and then arrange in order of boiling points and freezing points.

    a) 1.0 M $C_{12}H_{22}O_{11}$ ----> _____ ( ) highest b.p._____ highest f.p._____

    b) 3.0 M $Ca(NO_3)_2$ ----> _____ ( ) + _____ ( )   _____   _____

    c) 1.0 M $CaBr_2$ ----> _____ ( ) + _____ ( )   _____   _____

    d) 2.5 M $AlK(SO_4)_2$ ----> ____ ( ) + ____ ( ) + _____ ( ) _____   _____

    e) 2.5 M KCl----> _____ ( ) + _____ ( ) lowest b.p. _____ lowest f.p. _____

4. Using your knowledge of intermolecular bond strength, explain why polar solutes have less of an effect on the boiling point of water than ionic solutes.

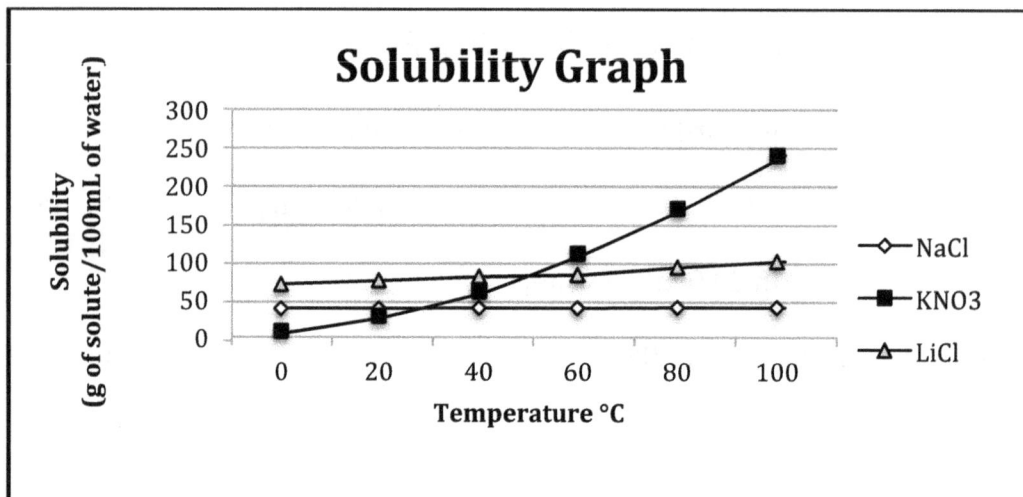

5. Describe what the solubility graph above tells you about the temperature and the solubility of the three ionic compounds.

6. The lattice energies of NaCl, LiCl, and KNO_3 are 787 kJ/mol, 853 kJ/mol, and 640 kJ/mole respectively. Use this information to explain why some compounds are more soluble in water by listing the bonds that are broken and formed during the dissolving process. Think carefully about the energy that is involved in bond breaking and formation.

7. What are three ways to increase the rate of dissolving of a polar solid in water and what do each of these methods do to the molecular interactions?

Name_____

Review Chapter 8

1. Show the solubility equation and then arrange in order of boiling point and freezing point.

a) _____ ---→ _____ ( ) + _____ ( ) highest b.p._____ highest f.p. _____
2.0 M potassium hydroxide

b) _____ ---→ _____ ( ) + _____ ( )   _____   _____
1.0 M lithium nitrate

c) _____ ---→ _____ ( )   _____   _____
3.0 M ethanol

d) _____ ---→ _____ ( ) + _____ ( ) lowest b.p. _____   lowest f.p. _____
1.5 M aluminum bromide

2. Fill in the states of matter on the following diagram. Include solid, liquid, gas, melting point and boiling point.

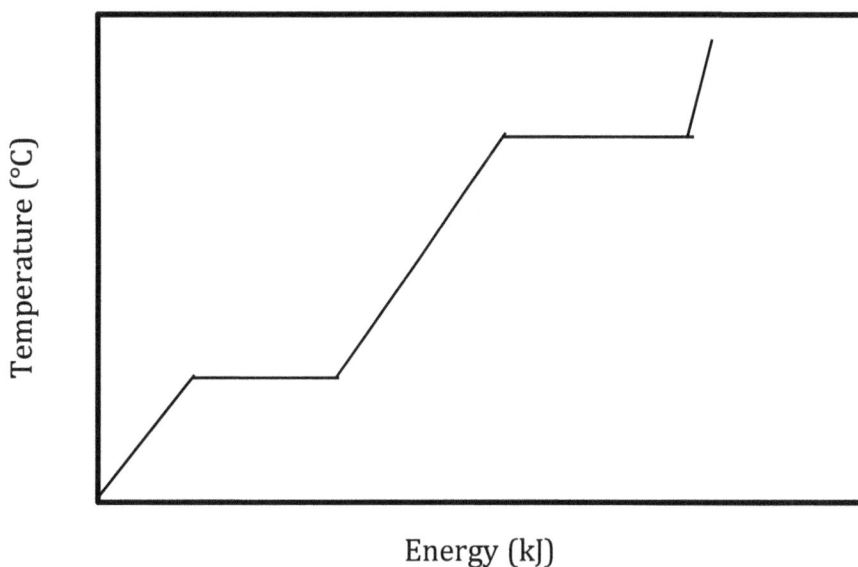

Energy (kJ)

3. Mark D if the relationship is direct, I if it is indirect, and N if there is no relationship.

a) ____ pressure and solubility of a gas in water (such as CO_2 in beverages)

b) ____ temperature and solubility of a gas in water

c) ____ temperature and solubility of a polar solid in water

d) ____ molecular mass and strength of dispersion forces

e) ____ melting point and strength of intermolecular forces

f) ____ volatility and strength of intermolecular forces

Appendix

Table of Useful Formulas   126

Table of Useful Formulas

Title	Formula
% Error	$\frac{|accepted\ value - experimental\ value|}{accepted\ value} \times 100$
Density	$D = m/v$ Density = mass / volume
Kelvin	$°C + 273$
Heat Capacity	$q = m \bullet C \bullet \Delta T$ heat = mass • specific heat • change in temp.
Coulomb's Law (electromagnetic force)	$F = kq_1q_2/d^2$
Mole	6.022×10^{23}
Ideal Gas Law	$PV=nRT$ R=0.0821 L•atm/mol•K
STP (standard temperature and pressure)	T = 0 °C or 273 K P = 1 atm or 760 mm Hg
Wavelength(m)	$\lambda = c / \nu$ $c = 3.00 \times 10^8$ m/s wavelength = speed of light / frequency
Frequency (sec⁻¹)	$\nu = c / \lambda$ $c = 3.00 \times 10^8$ m/s frequency = speed of light / wavelength
Energy (J)	$E = h\nu$ $h = 6.626 \times 10^{-34}$ J•s
M (molarity)	$\frac{moles}{Liter}$
Calculation for making a solution from a solid	M x V = moles (volume must be in Liters)
Calculation for performing a dilution	$M_1 \times V_1 = M_2 \times V_2$
% Yield	(experimental yield/theoretical yield)x100%
pH	$-\log [H_3O^+]$
pOH	$-\log [OH^-]$
$[H_3O^+]$	10^{-pH}
$[OH^-]$	10^{-pOH}
$K_w = 1.0 \times 10^{-14}$	$[H_3O^+][OH^-] = 1.0 \times 10^{-14}$
	pH + pOH = 14

www.ingramcontent.com/pod-product-compliance
Lightning Source LLC
Chambersburg PA
CBHW081545220326

41598CB00036B/6572

* 9 7 8 0 6 9 2 9 3 3 5 0 3 *